MONGE'S LEGACY
OF DESCRIPTIVE AND DIFFERENTIAL GEOMETRY

KRISTEN R. SCHRECK

DOCENT PRESS
Boston, Massachusetts, USA
www.docentpress.com

Docent Press publishes books in the history of mathematics and computing about interesting people and intriguing ideas. The histories are told at many levels of detail and depth that can be explored at leisure by the general reader.

Cover design by Nathaniel Parker Raymond, Graphic Details, Inc.

© Kristen R. Schreck 1999, 2016

All rights reserved. No part of this book may be reproduced or utilized in any form or by any means, electronic or mechanical, including photocopying and recording, or by any information storage and retrieval system, without permission in writing from the author.

Contents

Acknowledgements   **xiii**

1 Introduction   **1**

2 The Influences of Mathematics and Teaching on Monge's Life   **7**
    2.1 Monge's Early Years (1746–1769) 7
    2.2 Monge's Career in Teaching 15
    2.3 Monge at Mézières 17
    2.4 The Metric System 20
    2.5 École Polytechnique and École Normale 22
    2.6 Monge the Professor 30
    2.7 Monge's Mathematical Career 33

3 Monge's Mathematics   **37**
    3.1 Descriptive Geometry 37
       3.1.1 The Mongean Method 38
       3.1.2 Examples from *Géométrie Descriptive* . . . 41
               Cutting Planes Method 43
               Cutting Spheres Method 48
       3.1.3 Monge, Théodore Olivier, and String Models 52
       3.1.4 Descriptive Geometry After Monge 54
    3.2 Differential Geometry 58

	3.2.1	Examples from *Application de l'Analyse à la Géométrie*	61
	3.2.2	Involutes, Evolutes and Radius of Curvature of Space Curves	61
	3.2.3	Maple Exposition of Space Curves	83
		Helix	84
		Twisted Cubic	98
	3.2.4	Viviani's Curve	102
	3.2.5	Historical Perspective on Involutes	106
3.3	Monge's "Two Curvatures of a Curved Surface" . .	109
3.4	Maple Exposition of Curvature	113
3.5	Lines of Curvature	118
3.6	Historical Perspective on Curvature of Surfaces . .	120
3.7	Maple Exposition of Ruled and Developable Surfaces	124
	3.7.1	Cylinders, Cones, Hyperboloids of Revolution	125
	3.7.2	Developable Surfaces and Gaussian Curvature	137
	3.7.3	Minimal Surfaces and Mean Curvature . . .	146
3.8	Monge's Spirit in Differential Geometry	154

4 The Monge Tradition   **157**
4.1 The Advancement of Monge's Mathematics 157
    4.1.1 Charles Dupin (1784–1873) 158
    4.1.2 Jean-Victor Poncelet (1788–1867) 159
4.2 Monge's Influence in the United States 160
4.3 Conclusion 161

A The Mathematical and Scientific Works of Gaspard Monge   **167**

B Courses of Study and Academic Staff at the École
   Polytechnique (1794–1825)   171
   B.1 Analysis and Mechanics
        & Descriptive Geometry 171
   B.2 Chemistry and Physics   172
   B.3 Architecture and Fortifications 172
   B.4 Entrance Examiners   173

C Translation of Title Page and *Programme*
   from *Géométrie Descriptive*
   by Gaspard Monge, Paris, 1799   175

D Letters and Documents of Gaspard Monge and his
   family from the D.E. Smith Historical and
   Gouverneur Morris Collections (1772–1832)   181

E Three Translated Letters   187
   E.1 Translated Letter (Figure D.1) 187
   E.2 Translated Letter (Figure D.2) 188
   E.3 Translated Letter (Figure D.3) 188

Bibliography   191

List of Figures

2.1	Eighteenth century fortification drawing of Mézières	12
2.2	*Géométrie Descriptive* (1799)	27
2.3	Cylinder and cone as developable surfaces	29
3.1	Monge's construction of double orthographic projection from *Géométrie Descriptive* (1799)	40
3.2	Monge's cutting planes method–two cones	46
3.3	Monge's cutting planes method–conic section . .	48
3.4	Monge's cutting spheres method	50
3.5	Olivier string model of a ruled cylinder at the Musée des Arts et Métiers	53
3.6	Title Page of *Application de l'Analyse à la Géométrie* (1807)	60
3.7	Monge's construction of a plane curve	63
3.8	Circle and an involute	69
3.9	Circle and three involutes	69
3.10	Constructing the evolute of a plane curve	71
3.11	Parabola and its evolute	74
3.12	Monge's construction of an evolute of a space curve	75
3.13	Construction of an evolute to a space curve . . .	78
3.14	Helix with its involute	85
3.15	Helix with an evolute.	88
3.16	Helix with an evolute and three osculating circles	92
3.17	Helix with an evolute and three osculating spheres	95

3.18	Helix with an evolute as the edge of regression of the polar developable surface	97
3.19	Twisted cubic with an evolute lying on the polar developable surface	98
3.20	Twisted cubic with evolute as edge of regression of the polar developable surface	101
3.21	Twisted cubic on osculating sphere with evolute passing through its center	101
3.22	Viviani's Curve	103
3.23	Evolute and osculating sphere to Viviani's curve	105
3.24	Polar developable surface and osculating sphere to Viviani's curve	106
3.25	Normal section of saddle surface with $\kappa_1 > 0$	116
3.26	Normal section of saddle surface with $\kappa_1 = \kappa_2 = 0$	117
3.27	Normal section of saddle surface with $\kappa_2 < 0$	118
3.28	Lines of curvature on an ellipsoid from *Application de l'Analyse à la Géométrie* (1807)	119
3.29	Ruled cylinder	127
3.30	Ruled cone	128
3.31	Hyperboloid of revolution	129
3.32	Ruled hyperboloid of revolution with two generators highlighted	131
3.33	Ruled cylinder with figure eight base curve	131
3.34	Ruled cone with figure eight base curve	134
3.35	Right conoid with two rulings highlighted	134
3.36	Tangent developable surface of a helix	140
3.37	Tangent developable surface to Viviani's curve	140
3.38	Tangent developable surface to a helix shaded by its Gaussian curvature: $K = 0$	141
3.39	Hyperboloid of revolution shaded by Gaussian curvature: $K \neq 0$	146
3.40	Hyperboloid of revolution shaded by its mean curvature: $H \neq 0$	149

3.41	Catenoid shaded by mean curvature: $H = 0$	152
3.42	Enneper's surface shaded by mean curvature: $H = 0$.	154
4.1	Monge's Grave at Père Lachaise Cemetery in Paris (front view)	165
4.2	Monge's Grave at Père Lachaise Cemetery in Paris (back view)	166
D.1	Letter from Monge in his position as the Minister of the Marine to the Commissary of the National Treasury (1793)	182
D.2	Letter to Students of the École Polytechnique from the Jury of Examiners: Bossut, Laplace, Monge, and others (1799)	183
D.3	Letter from Monge in his role as the Minister of the Marine regarding hydrographic testing (1790)	184
D.4	Document regarding the Centennial plans for the École Polytechnique (1895)	185
D.5	Monge Portrait from Géométrie Descriptive (1799)	186

Maple Code Listings

3.1	Circle with one involute	66
3.2	Circle with several involutes	67
3.3	Involute of a curve	68
3.4	Evolute of a curve	73
3.5	Involute of a helix	85
3.6	Evolute of a helix	86
3.7	Evolute of a curve	87
3.8	Three osculating circles	91
3.9	Three osculating spheres	94
3.10	Polar developable surface	96
3.11	Evolute and polar developable surface of twisted cubic.	99
3.12	Twisted cubic with osculating sphere	100
3.13	Viviani's curve	103
3.14	Evolute of Viviani's curve	104
3.15	Polar developable surface to Viviani's curve . . .	105
3.16	Animation of normal sections on saddle surface .	115
3.17	Ruled surfaces: cylinder, cone and hyperboloid of revolution	126
3.18	Generators of hyperboloid of revolution	132
3.19	Modified ruled cylinder and cone	133
3.20	Generators of right conoid	136
3.21	Tangent developable surfaces	143
3.22	Computation of Gaussian curvature, K	144

3.23	Coloring surfaces by Gaussian curvature, K	145
3.24	Computation of mean curvature, H	150
3.25	Mean curvature: hyperboloid of revolution	151
3.26	Mean curvature: catenoid	152
3.27	Mean curvature: Enneper's surface	153

Acknowledgements

In the midst of pursuing my Doctorate of Arts degree at the University of Illinois at Chicago, I took a course in differential geometry of curves and surfaces in three dimensions with my advisor, Steven Hurder. Steve made the subject come alive as we designed thrilling roller coasters and imagined ourselves as tiny insects traversing wildly twisted, curving surfaces. I was enthralled with the subject and wanted to learn more about the nature of this intriguing area of mathematics. Steve suggested I start by finding the origins of the history of differential geometry. It was then that I discovered the mathematical works and fascinating life of Gaspard Monge, the "Comte de Péluse". Studying Monge became the focus of my academic work during my life as a graduate student and his mathematics became the essence of my dissertation.

I wish to thank Steve for being a most fantastic and dynamic thesis advisor and teacher. Steve is a truly caring and giving advisor and also an exceptional person. Our shared enthusiasm for differential geometry and love of France will always keep us connected. I also wish to thank Peg Bogacz, she has been my faithful mentor and dear friend who motivated me to pursue the Doctorate of Arts degree. Peg has been an inspiration to me ever since. I thank the other the members of my dissertation committee, John Wood, William Howard, and Philip Wagreich (1941–2013) for their thoughtful suggestions and genuine interest in my subject when I first began writing about Monge. I also wish to thank my loving parents, Raymond and Georgia Anderson, for their ever present encouragement as I continued my education over the years and for showing me the importance of being well rounded in life.

Most importantly, I am grateful to my sweet husband, Deron, and my darlings, Isabella and William, for their love and eagerness to learn all about Monge. It was a delight to witness their curiosity as they shared in the excitement of this new endeavor with me.

I wish to express my gratitude to Mary Cronin, who initially contacted me about writing a book based on my dissertation, she helped to make a dream of mine come true. I would also like to express my appreciation to Paul Hamburg for his extraordinary skills in translating the original 18th century French documents of Monge's *Programme* from *Géométrie Descriptive* (1799) and some of Monge's personal and business related letters written in his own hand. I am appreciative to David Stern, Susan Klaiber, and Martha Pollak for their expertise and guidance in finding graphic material on 18th century French fortifications. I would like to acknowledge the hospitality of the Newberry Library and Columbia University. Finally, I thank my editor, Scott Guthery for his cheerfulness and endless words of encouragement that kept me focused and energized. I could not have completed this book without Scott's sound historical, mathematical, and technical advice and expertise.

Chapter 1

Introduction

> Gaspard Monge was "perhaps the most influential mathematics teacher since the days of Plato"[1]
>
> Carl Boyer

A gifted mathematician and expert geometer, Gaspard Monge (1746–1818) was an inspiring educator with an innate ability to spark a fire within his students. Outside of mathematics, he was adept at cultivating political alliances with highly influential friends. Monge's everlasting contributions to mathematics were the development of descriptive geometry and the founding of the field of differential geometry.

Monge's descriptive geometry was created during his early years (1768–1769) while he was attending military school at the École Royale du Génie de Mézières. It was based on the pure geometric methods of projection, perspective, and cross section. These methods originated from the functional requirements of designers and draftsmen to represent three-dimensional objects on a two-dimensional plane when constructing plans of fortifications. In contrast, Monge's work in differential geometry was theoretical,

---
[1] Boyer, 1960, p. 13.

comingling analysis and geometry and it was aimed at mathematicians of the learned world. His first lectures on this subject, *Feuilles d'analyse appliquée à la géométrie,* were given in 1795 to students at the École Polytechnique in Paris. In his course at the École Polytechnique, Monge explored applications of calculus to the study of curves and surfaces in three dimensions.

In this book, the historical evolution of Monge's mathematics and the practical techniques he used in creating and teaching descriptive and differential geometry during the late eighteenth century in France are examined. Examples from Monge's original mathematical writings reveal the revolutionary methods he set forth during this time period and illustrate his effectiveness as a teacher. It is from these examples where one gleans Monge's legacy and its impact on the teaching of mathematics today. The computational and graphical capabilities of the computer algebra system, Maple, are utilized to bring Monge's ground-breaking ideas to life and into the twenty-first century. Furthermore, the captivating individual Monge was and the fascinating life he led are uncovered.

Monge's mathematical career began during his early years of schooling. His aptitude for accurately describing complicated figures in space by using words and equations, in conjunction with his skillfulness in visualizing those figures with ease, by constructing drawings and physical models, made him a natural in geometry. When Monge was still in his teens, his mathematical ability and scientific brilliance were recognized by his teachers, influential scholars of science, and important members of government. These acknowledgments shaped the course of his lifelong contributions to mathematics and his dedication to perfecting the methods of mathematical instruction.

To understand Monge's genius as a mathematician and as an educator, it is worthwhile to know the mathematical climate in France during the late eighteenth century. Calculus and analytic geometry were the mathematical creations of the seventeenth century. The nineteenth century gave "rise to mathematical rigor and the flowering of geometry".[2] For comparable mathematical developments toward the end of the eighteenth century, one may look especially to France during a critical time in its history. The mathematicians in France at the time of the Revolution (1789) were unique because of their vigorous interest in creating new and challenging mathematics. Their theories were further developed by mathematicians of the nineteenth century and then expanded across the globe. It might be said that two other revolutions occurred within the mathematical community at this time in Europe, particularly in France, a geometrical revolution and an analytical revolution.[3]

Joseph-Louis Lagrange (1736–1813), Pierre-Simon Laplace (1749–1827), and Adrien-Marie Legendre (1752–1833) were some of the best known mathematicians in France during the time of this productive generation of mathematics. However, it is Monge who is credited for the revival of the study of geometry, celebrated for his strong mathematical insights, and lauded for his leading role in the teaching and the proliferation of technical education throughout France.

In this twenty-first century, if someone is exceptionally talented in mathematics and has the desire to work in a related field, we might assume that individual will find a suitable position in academia or industry. During the eighteenth century in France, universities were not the centers of mathematical research as they are known to be today. The Universities at Paris and Oxford were the leading mathematical hubs during the fourteenth century. However, by the time of the eighteenth century, Paris had lost its position because of France's adherence to outmoded mathematical methods.[4] Consequently, the universities in France were no

---

[2] *Ibid.*, p. 11.

[3] *Ibid.*, p. 12.

[4] *Ibid.*, p. 12. For instance, as a large part of the world's scientific community embraced the mathematical ideas of Isaac Newton (1643–1727) and

longer centers that provided advanced education in mathematics. For example, there were no leading mathematicians even associated with the University of Paris. French mathematicians could cultivate their profession only by the support of private sponsorship, the church, or the military. This resulted in the noteworthy mathematicians of the time being associated with various academies created by royal patronage. These academies were founded to contribute to the monarch's prestige by providing mathematical and scientific education, which in turn, produced military engineers for the advancement of the nation.

At the end of the eighteenth century, the revolutionary government appointed Monge, Laplace, and Lagrange to develop and document proven methods for teaching in their respective fields. Motivated by the need for centralized instruction in the preparation of teachers and training of military engineers, two schools of higher education were formed. A new, contemporary school of its time, the École Normale was formed in 1794 to produce future teachers and administrators. It is where Laplace taught and also where Monge gave his first public lectures showcasing his methods in descriptive geometry. The school was discontinued after a short time, it was reorganized and re-established as the École Normale Supérieure years later.

Monge led this group of mathematicians, as director, to establish the École Polytechnique of Paris in 1794. The school's curriculum emphasized mathematics, both theoretical and applied, as an integral part of the education and training of military engineers. The lectures Monge presented on curves and surfaces at the École Polytechnique became the first exposition in the area of differential geometry. Many of Monge's students were left with enthusiastic

---

Gottfried Leibniz (1646–1716), the mathematicians in France during this time held fast to the Cartesian methods (according to Boyer, 1960, p. 13), and were not readily willing to accept the new ways. In effect, France had consistently become a step behind.

memories of his geometry lessons. Monge worked hard and proved to be an effective teacher in training his students to undertake new research on the road he himself followed. The influence of his mathematics and teaching methods even extended to the United States. In 1816, Mongean descriptive geometry was introduced into the curriculum at the United States Military Academy at West Point by a former student of the École Polytechnique, Claude Crozet (1790–1864). Thus, France's impact in regards to the rich mathematical curriculum Monge instituted, was heartily kept alive.

Chapter 2

The Influences of Mathematics and Teaching on Monge's Life

2.1 Monge's Early Years (1746–1769)

The experiences Monge enjoyed early on in life hold the key to understanding his mathematical and educational impact in France, and, eventually, throughout the world. Gaspard Monge was the eldest son of merchant, Jacques Monge (1718–1775) originally from Haute-Savoie, and, Jeanne Rousseaux (1711–1781) from the region of Burgundy. He was born on May 9, 1746 in Beaune, France. Monge had two brothers, Louis (1748–1827) and Jean-Baptiste (1751–1813). All three Monge brothers possessed mathematical ability and each earned a reputation of his own. Louis Monge was an astronomer, examiner in hydrography for the navy, and a professor of mathematics at the École Royale Militaire in Paris and also at the École Royale du Génie de Mézières. The youngest son, Jean-Baptiste Monge was a professor of mathematics at the École

Militaire in Rebais and later became a professor of hydrography in Antwerp, Belgium.

To Jacques Monge, educating his sons was paramount. He toiled diligently as a local businessman to give his sons a solid education and sought to send them to schools of the highest quality. All three brothers received their first education at the local college of the Congrégation de L'Oratoire, directed by a religious order in Beaune. The school provided a twelve year education in the primary and secondary grades. Monge was an exceptional student with superior knowledge and special ability in mathematics and the sciences. He routinely earned the highest grade in his classes. He graduated "puer aureus," the golden boy, of his year in 1762 after giving his thesis defense (in Latin) to a public audience on the elements of "calculation and of geometry."[1] It was clear that Monge was a born geometer and engineer with an unsurpassed gift of visualizing complicated space-relations.[2]

Monge studied at the Collège de la Trinité at Lyon from 1762 to 1764. In Lyon, at the age of sixteen, he was given the chair of the physics department because of his instinctive knowledge and advanced analytic and scientific skills. He was an affable and patient young man who lacked any affectation. These traits, combined with his sound knowledge, contributed in making him a great teacher. After completing his education in Lyon in 1764, at the age of eighteen, Monge returned to Beaune and, remarkably, from that point forward he did not have any further formal scientific training.

In the summer of 1764, in his leisure time, Monge sketched a large-scale topographical and architectural map of the city of Beaune. Monge had no prior training in this type of drafting and he used rudimentary equipment of his own construction. His superb drawing of the city attracted the attention of a longtime teacher at the École Royale du Génie de Mézières, Colonel Du Vignau of

---

[1] Gillispie, 1980, p. 522.
[2] Bell, 1945, p. 184.

the French engineers. This map gave Monge the first significant opportunity for his knowledge and talents to be recognized by his superiors. Indeed, Monge's draftsmanship so impressed Vignau that he offered Monge the position of draftsman at the military school Mézières. Because of his parents' humble background, he was not considered for an officer's position. Monge accepted the position of draftsman initially with some disappointment. However, he quickly took to the tasks of preparing plans for fortifications and architectural models, stone-cutting, and carpentry. All the while, Monge waited for the opportunity to demonstrate his mathematical expertise. In hindsight, the position as draftsman played a most important role in the direction of his career.

The military school of the École Royale du Génie de Mézières was founded in 1748 for the education of engineering officers. Mézières trained 452 military engineers in the years before the Revolution of 1789.[3] It was a prestigious school, which offered high quality practical and scientific training through a two-year course of study. Cadets were trained primarily in the construction of fortifications, and were also called upon to design bridges, windmills, warehouses, and churches. The school rapidly earned an outstanding reputation for producing talented recruits. The size of the academy fluctuated from year to year based on the estimated number of engineers that were needed at the time. The selection process of the candidates was based on intellectual as well as social criteria.[4] It was rigorous and open to everyone, regardless of class.

---

[3]Gillispie, 1980, p. 511. This was the number of officers who met the requirements set by the examiners during the time period. From 1758 to 1776 enrollments averaged 30 cadets every four years. No students were admitted from 1772 to 1788 to lower the enrollment.

[4]Gillispie, 1980, pp. 512–514. In the first stage, the candidates had to cross a social threshold before being given a competency examination. The candidate was required to submit a birth certificate, his family status or circumstances, and evidence of future financial support if he were to be admitted to the corps. In the second stage, the candidate with the pedigree still had to pass the

Renowned teachers at the school were Monge and Charles Bossut (1730–1814), and among the school's distinguished students were Jean-Charles Borda (1733–1799), Lazare Carnot (1753–1823), and Joseph Bertrand (1822–1900). France was known for its mathematical excellence in the application of analysis to mechanics at this time. This system of analysis was taught in the technical schools and directed the course of mathematics during the Revolution. During these same years, England was a leader in synthetic geometry.[5] The main disparity between England and France in this age focused on the need for stronger technical training in French schools of higher learning.

In 1768, at the age of 22, Monge's shining mathematical moment arrived at Mézières. He was asked by one of his supervisors to solve a practical problem for defilading a fortification using data provided by certain observations.[6] To defilade a fortification means to erect the walls of the fortification in such a way so as to shield crucial positions from both the view and fire power of the enemy no matter where the enemy was located. Fortifications varied from country to country. For example, in England fortification was really only an academic study as the English Channel made an effective natural barrier. However, in France, the land boundaries

---

entrance examination which consisted of elements of arithmetic, theoretical and practical geometry, and statics.

[5]Daston, 1986, p. 269. Synthetic geometry in the late-eighteenth century explored the relationship between mathematics and the physical sciences. Synthetic geometer's strove to widen the scope of traditional geometric methods by introducing what were called explicitly physicalist concepts such as transformation, projection, and continuity. Mechanically-based techniques were used to find geometric analogues of analytical entities such as infinitesimals, negative numbers, and imaginary numbers. Work in synthetic geometry was viewed as a meticulous alternative to the analytic routes exemplified by eighteenth century French mathematics.

[6]Booker, 1963, p. 18. In the 17th and 18th centuries, fortification in France was based on the bastioned system used by the heralded military engineer Sébastien Le Prestre de Vauban (1633–1707).

and frequent wars made fortification a crucial, precise science. The fortifications were geometrical in shape, often based on a hexagon or pentagon, which grew outward at various higher places for musket cross-fire. Usually the inner fortress of the fortification was surrounded by a large ditch or moat encircled by complicated barriers that radiated beyond. Sébastien Le Prestre de Vauban (1633–1707), engineer to Louis XIV, brilliantly adapted fortification design techniques created in the 16th century to construct fortresses that could withstand various siege operations. His methods became the classical way to construct fortifications in France.

The established method of defilading a fortification in the late eighteenth century involved many calculations based on empirical data obtained from measured observations. These calculations followed a strict criterion and inherently had a large and costly safety margin. Monge devised a new approach using purely geometrical constructions to solve every problem efficiently and significantly faster than was expected. By convention, there was an expectation that a lengthy amount of time was required to create these drawings. As a result, the officer in charge did not accept Monge's drawings at first, until he examined them more closely.

In his solution to the problem, Monge extended plane geometry into three dimensions. If the fortification was constructed on a flat terrain, defilading presented no difficulty. With uneven terrain, the problem was more intricate. Monge's method consisted of constructing a site plane, that is, a plane tangent to a wedge or cone shaped surface. This point on the fort would have been selected based on minimum earth-moving conditions.[7] The base curve of the cone represented the highest points of the fort that can be seen from the vertex. The site plane would be taken as a basis upon which the walls of the fortification would be built similar to a flat terrain, instead of the natural horizon plane. Monge's technique projected the necessary points onto this reference site plane, drawn

---

[7]Monge, 1781.

in elevation.[8] An example of the fortification at Mézières is given in Figure 2.1.

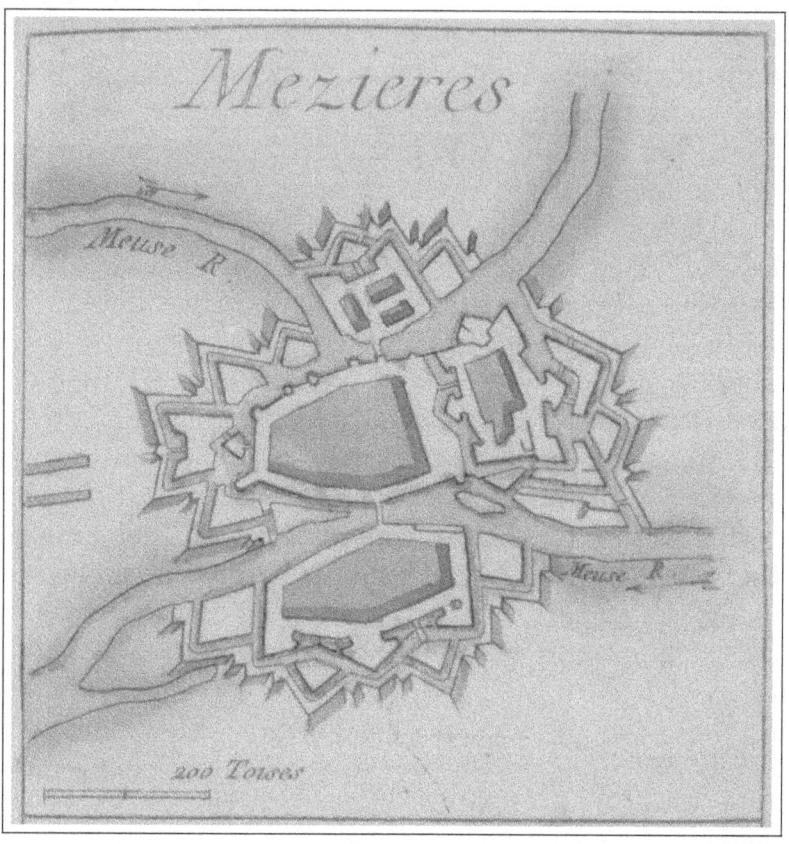

Figure 2.1: Eighteenth century fortification drawing of Mézières[9]

---

[8]Belhost, 1996, p. 53. A clear depiction of the method Monge would have used in constructing a site plane is given in the detailed drawing indicated on the referenced page of Belhost. Also, see Figure 9.5 in Langins, 2004, p. 248 for an illustration showing Monge's method of defilading.

[9]Le Rouge, 1755, p. 27. Photo courtesy of the Newberry Library, Chicago. Vault Case G1838.L47.

Since this astounding work was now purely geometrical, the constructions could be performed more expeditiously with a clear visual record of the work that was done. This was quite in contrast to the map-making techniques that relied on surveying and tedious calculations which were used prior to Monge's developments. With previous methods, a problematic construction issue may have required structures to be torn down and rebuilt. Monge's approach was unique in that he generalized his drawing practices so that a draftsman could see the underlying principles involved. His methods enabled the designer to replicate and apply similar techniques to other problems of the same nature. The representations produced using Monge's process made it easy to extract pertinent information from the drawings, allowing questions with the architectural or mechanical design to be readily resolved. With Monge's geometrically based system, the unsolved problems in the past became simple tasks. This was the beginning of descriptive geometry.

Monge's training had led him to broaden his new method for designing fortifications and apply it to other kinds of technical drawings. His work provided clarification of the essential axioms of mechanical and architectural design and brought together perspective, the theory of shadows, and cartography, using Euclidean and Cartesian geometry. This doctrine, as Monge himself termed his collection of axioms, elevated descriptive geometry to mathematical prominence. The simplicity and clarity of his methods of descriptive geometry revolutionized mechanical design and became the basis for industrial design as it is currently known. Due to Monge, descriptive geometry was legendary and it became the "new Mathematics of its day."[10]

Monge was generously recognized for his new geometrical methods in designing fortifications and was awarded a promotion at Mézières. He became *repetiteur,* or tutor, under mathematics pro-

---

[10]Grattan-Guiness, 1981. p. 362.

fessor Charles Bossut[11] and the part-time assistant to the chair of the physics department, the Abbé Antoine Nollet. [12] A year later in 1769, Monge succeeded Bossut (although he was not given the title of professor), on the understanding, as the story has been told, that the results of his descriptive geometry were to remain classified because of its uses in the military. It was stipulated that the subject could only be taught at Mézières which, of course, would bring a certain amount of prestige to the military school. Monge continued his study into the nature of descriptive geometry and methodically constructed a course on the subject matter at Mézières. However, Monge's aspirations to spread his new doctrine throughout France were hindered due to the ban on the publishing of his methods of descriptive geometry.[13] To overcome his disappointment, Monge began to work in another direction. He pursued the theory of curves and surfaces in three dimensions by applying geometrical reasoning to applications of calculus, the implementation of which was known to be characteristic of his manner of thinking.

---

[11]Taton, 1951, p. 11. Bossut (1730–1814) was the chief examiner at Mézières after 1768. He was a professor of mathematics and mechanics at the military academy for fifteen years. In 1772, Bossut wrote two volumes of the mathematical textbook *Histoire générale des Mathématiques,* published in 1802 and 1810 for use by the students at the school.

[12]Gillispie, 1980, p. 506. The Abbé Antoine Nollet (1700–1770) was a celebrated physics instructor at Mézières and was considered to be the "foremost electrical theorist of the mid-18th century".

[13]Booker, 1963, p. 21. The ban on the widespread distribution of Monge's *Descriptive Geometrie* was not lifted until 1794 and his monograph on the subject was not published until 1799. This thirty-year postponement may have been beneficial in that Monge's final work was more mature and was received by a different audience. According to Gillispie, 1980, p. 523, this account may be a bit of an exaggeration. The methods for defilading a fortification at Mezieres were confidential, however, the documentation that confirms Monge was prevented from publishing his discoveries in descriptive geometry remains to be seen.

2.2 Monge's Career in Teaching (1769–1795)

Monge's illustrious teaching career paralleled his mathematical and scientific work. When he began teaching in 1769 at the military school at Mézières, he taught the "select subject unique to Mézières,"[14] a creation of his own vision, descriptive geometry. It was there that Monge gained teaching experience in the scientific education of engineers in which mathematics played the highest role. Monge proved that he was a gifted mathematician, a competent physicist and able experimenter, a skilled draftsman, and a first-rate teacher. However, Monge understood that he was not likely to be recognized as a mathematical leader simply by teaching at Mézières, so he often spent his holidays in Paris conferring with the leaders in science. Monge desired to become part of the "learned world."[15] He strove to create a name for himself, so that his mathematical works would be discussed in the elite circles.

His ambition led him in 1771 to meet the marquis de Condorcet[16] who had been a member of the Académie des Sciences in Paris since 1769. The Académie des Sciences had been established in the seventeenth century as an outgrowth of a meeting arranged by Father Marin Mersenne.[17] Father Mersenne developed a

---

[14]*Ibid.*, p. 21.

[15]*Ibid.*, p. 21.

[16]Boyer, 1960, p. 20. Marie-Jean-Antoine-Nicolas Caritat, marquis de Condorcet (1743–1794) was an accomplished mathematician who had published books on integral calculus and the application of probability and statistics to social problems. He was an ardent political activist, especially on those issues related to the welfare of mankind and social injustice. He played a large part in making the Revolution happen. Interestingly, Condorcet held the belief that "education would eliminate vice" and he fervently argued for free education.

[17]Caullery, 1933, p. 14. Mersenne (1588–1648) was a Minimite friar and close friend of Descartes. Some initial members of the Académie des Sciences were Fermat, Descartes, Pascal, and Desargues.

network of communication and organized regular meetings among a diverse group of mathematicians and scientists even though he himself was not a man of science. He created a bond between the scholars of science who were living in Paris, in the provinces of France, and in other countries. After Father Mersenne's death, the meetings continued and Jean-Baptiste Colbert[18] was responsible for organizing what evolved into the formation of the Académie des Sciences in 1666. In 1771, Monge presented four papers[19] to the Académie des Sciences which highlighted the main scope of his research:

1. "Les équations de maximum qui contiennent des différences premières" (March 6, 1771). This first paper broadened the study of the calculus of variations to the study of extrema and double integrals. Monge's work in this area began and ended with this paper.

2. "Les développées des courbes à double courbure et leurs inflexions" (August, 31, 1771). Considered to be Monge's most original work on the equations of curves and surfaces in three-dimensions, it was not published until 1785. Monge's essays during this span of time presented a large part of the new and avant-garde theories that he further developed in his later papers. They also provided a clear example of his teaching style and "his very personal method of exposition, which combined pure geometry analytic geometry and infinitesimal calculus"[20]

---

[18]Colbert (1619–1683) was an administrator and politician under Louis XIV. Colbert was the man behind the king in making the initiatives and administrative decisions in the New France during the years 1661–1672. The French called the territories that they took possession of starting in the 17th century "the New France".

[19]Taton, 1851. Also see Appendix A for a chronology of Monge's mathematical and scientific publications.

[20]*Dictionary of Scientific Biography.* Vol. IX, 1974, p. 470.

3. "L'Intégration des équations aux différences partielles" (November 27, 1771). The general idea of associating partial differential equations with geometry in three dimensions, as developed in this paper, had weighty influence in the evolution of differential geometry of curves and surfaces. Monge furthered this work in later years in four publications dating from 1776.

4. "Un tour de cartes" (December 20, 1771). Monge's fourth paper involved card tricks using combinatorics.

2.3 Monge at Mézières and as Naval Examiner

Monge was deeply committed to the future engineering careers of his students at Mézières. He aimed to do all he could to afford his students a solid technical education and to instill in them the value of science. In 1775, Monge was recognized for his exemplary teaching and awarded the formal title of "royal professor of mathematics and physics" by the academy's administration.[21] Monge's varied mathematical investigations and his innovative methods exhibited many similar characteristics. He possessed strong geometrical intuition, a keen interest in practical problems, and profound analytical ability. To that end, Monge taught his students to examine a single problem from three aspects, the analytical, the geometrical and the practical.

Monge married Catherine Huart in 1777 and they had three daughters, Charlotte Emilie (Marey), Louise Francoise (Eschasseriaux), and Adelaide. During 1777–1780, Monge's concentration was primarily in chemistry and physics. He was responsible for creating a state-of-the-art chemistry lab at Mézières. In

---

[21] *Ibid.*, p. 470.

1780, at the age of 34, he was elected to the Académie des Sciences as "adjoint géomètre" in which he became a deputy surveyor-topographer. This changed his teaching life to a large extent. He was required to spend a considerable amount of time in Paris participating in projects and presenting his work in mathematics, physics, and chemistry. Monge divided his time between Paris and Mézières for a few years while keeping his position and salary at Mézières. During the time he was away in Paris, Monge's role as an educator was never far from his mind. He arranged for his brother Louis to cover his courses.[22] Monge always maintained an intense interest and concern for the military school that helped to form his own teaching style and approach. He kept a careful watch over the progress of its students, paying particular attention to the curriculum and research activities.

In 1783, the French government appointed Monge head examiner of candidates for the navy, replacing Étienne Bézout.[23] This was a significant appointment as it began Monge's future in public life in France. As chief naval examiner, he was in charge of directing a branch of France's military maritime schools and he regularly toured each institution under his administration. During his visits, he discovered a less than satisfactory level in the naval educational system. Monge was resolute in reforming the structure of the naval educational system to graduate well-trained officers. He instituted permanent improvements to technical and scientific instruction amidst maintaining high standards. He de-

---

[22]Gillispie, 1980, p. 528. Louis' regular position was at the École Militaire in Paris. Interestingly Napoleon Bonapart was one of Louis' students there.

[23]Taton, 1951, p. 26 and Gillispie, 1980, p. 506. Bézout (1730–1783) along with Bossut were two principal authors of mathematical textbooks at Mézières. Bézout wrote a multi-volume compendium the "Cours Complet de Mathématiques" which covered the entire spectrum of mathematical instruction from arithmetic through the applications of calculus. He was well known for his research on solving equations using determinants and the method of elimination.

voted six months out of the year to observe teaching instruction and conducting examinations of the young naval trainees. He became knowledgeable of the full naval establishment. During his travels, he also had the opportunity to examine the iron mines, factories, and foundries in each place he toured. Monge became an expert in metallurgy and manufacturing operations that were based on the newest technologies.

The extent of time required as naval examiner, together with his Académie des Sciences duties, made it necessary for Monge to resign his professorship at Mézières in 1784, the military academy at which he spent twenty successful years. In the course of the next eight years, 1784 to 1792, monitoring the naval schools throughout France and participating in the Académie des Sciences in Paris were Monge's primary activities. There was a blossoming of Monge's interest in the sciences, particularly in physics and chemistry, during these busy years as well. In 1785, Monge participated with Antoine Lavoisier[24] in experiments on the composition of water. In another experiment with Jean-François Clouet (1751–1801) they were able to successfully liquefy sulfur dioxide. He presented his work in these areas to the Académie des Sciences. In addition, Monge wrote an elementary textbook on statics, *Traité Élémentaire de Statique*, which was published in one volume in 1788. This textbook became a classic on the subject. An English translation by A. M. Woods Baker, *An Elementary Treatise on Statics by Gaspard Monge with a Biographical Notice of the Author*, was published in 1851.

---

[24]Booker, 1963, p. 23. In 1785 Monge had worked closely with Antoine Laurent Lavoisier (1745–1794). Lavoisier was not only an exceptional chemist, but he was also highly interested in education. Around 1791, as a member of the committee established by the National Assembly to advise the government on vocational careers, Lavoisier submitted a proposal for a national system of public education. Lavoisier's beliefs supported Monge's own ideas to create a new system of education. Consequently the École Normale was created in 1794. Sadly, Lavoisier did not have the opportunity to witness the fruition of his ideas, he was executed earlier that year.

2.4 The Metric System

In May of 1789, a meeting of the États Généraux took place at Versailles. Several weeks later the group became known as the National Assembly in France and the Revolution became a reality. The general upheaval was a scientific revolution as well as one against the government, and in this Monge was a leader.[25] Monge and Lagrange were two of the prime motivators in the establishment of the Metric System. They were members of a committee who devised a report on a new system for assigning conventions to weights and measures. Their task was to ensure that the new scheme was adopted by the general population. The report was eventually published in 1791 in the *Memoires* of the Académie des Sciences. From this committee emerged the Commission on Weights and Measures composed of Monge, Lagrange, Laplace, and Condorcet. Ultimately, the *metric system* materialized from this Commission.

As part of the Commission's work, Monge and his colleagues defined the meter as the standard of length based on their investigations of the distance along the terrestrial meridian.[26] The standardized measuring structure and decimal representations offered by the metric system made it especially befitting for scientific and engineering work. The new system made it possible for figures in industrial drawings to be represented by their dimensions. Thus, design and manufacturing processes could clearly be delineated. Rapid developments in technology were widespread and made possible by the adoption of the metric system. Monge's role in the

---

[25]Booker, 1963, p. 22. In 1774 in Britain, the Society for the Encouragement of Arts, Manufactures, and Commerce (more commonly known as Royal Society of Arts) had offered 100 guineas for "a mode of ascertaining invariable standards for weights and measures, communicable at all times to all nations".

[26]The meter was defined as one ten-millionth of the great circle distance between the equator and the North pole along the meridian passing through Paris.

formation of the metric system is of consequence here as each of his experiences throughout his career played a compelling part in the development of his unusual way of thinking. Every involvement in his one-of-a-kind undertakings: mathematical, scientific, educational or political ignited his genius and affinity for invention. Monge's abiding attitude to unveil the connections between theory and the real world is a theme throughout his life that had always prevailed.

From 1792 to 1794, Monge took on more active roles in politics and government. After the downfall of the monarchy in August of 1792, Monge accepted the position of Minister of the Marine. This position was offered first to Condorcet, but he refused the post and nominated Monge. Monge was a steadfast supporter of the Revolution and a devout Republican, but he never took part in any type of organized political action. Monge served earnestly in his new post and put forth all of his efforts as Minister of the Marine to assure France's existence and independence. The French naval fleet was in a terrible state, unorganized and ineffective. Politically, Monge was in a tough spot. He was in the middle of opposing political extremes with liberals on the one side and conservatives on the other side. Through his administrative agility, the French naval fleet was rejuvenated by his efficient methods. New provisions were set up to suppress contraband in the navy and Monge worked hard to maintain an *esprit de corps* with the men in his command.

This was at a time during the Revolution when the fight for the cause for liberty was at its peak. Monge energetically sought for governmental aid to pay salaries and other expenses. He was also in touch with the French colonies in America, sending out carpenters to serve under orders of the Minister of France to the United States of America. However, the troubles of the French government continued to increase, making Monge's political position in his job more and more intolerable. In April of 1793, Monge resigned as Minister of the Marine. It was a wise decision on his part, the political extremists began to threaten Monge and he left Paris in order to elude the guillotine. After a few months, Monge returned only to become one of the eager men selected by a minister of state to aid in the foundation of, and what grew to become, some of France's most glorious schools.

2.5 The Establishment of the École Polytechnique and the École Normale

Most important to the future of mathematics in France, especially after the crisis of the 1792 undoing of the King, was the endeavor Monge took on to institute a technical school for the tutelage of engineers.[27] In 1793, a new Committee of Public Safety was established. An aim of this committee was to attract scholars and appeal to them to use their talents by banding together in defense of the country. As a member of the subcommittee, the Committee on Arms, Monge was responsible for developing the manufacturing of arms and munitions. Even as France held an army of close to one million men to defend the nation, the factories were lagging so far behind technologically due to the lack of educationally well-prepared engineers in the workforce. They were unable to manufacture even one tenth of the materials that would be required to arm the corps.[28] Monge, the enthusiastic patriot, recognized two possibilities. First, industry could profit greatly from science. Second, Monge could prove to his scholarly friends that, together, they could apply their acumen and know-how to the manufacturing process. Specifically, the objectives were to "increase manufacturing

---

[27] Years before, in 1786, Monge was a chief proponent of teaching reform in the naval academies. His work in this regard was excellent preparation for his future efforts in the revitalization of scientific and technical education that he pursued during the Revolution.

[28] Booker, 1963, p. 22.

capacity, to simplify manufacturing processes, to produce the right metallic alloys; particularly to make steel in large quantities, and to provide a prodigious amount of gunpowder all in extraordinary haste".[29]

Relying on his vast experiences in traveling throughout France, Monge knew the right people who could assist in accomplishing the goals he set forth. He made numerous connections, forging friendships with journeymen, draftsmen, and designers, mathematicians and scientists, along with educators and administrators. His diverse encounters stood him well. The passionate efforts Monge put forth in defense his country, in conjunction with his reputation for having unwearied energy, distinguished him from other academics.

To solve these problems, Monge visited arms and munitions factories, forges, powder mills, and foundries. He carefully collected data on the techniques that were being utilized at each plant. He would then analyze his findings with the goal of simplifying and improving upon the operational processes that were used. Monge wrote several significant papers[30] from the information he gathered: "Saltpetre, from earth to cannon in three days," "The art of making cannons", and, "Advice to iron-workers on the fabrication of steel." In writing the last paper, he collaborated with his colleagues Berthollet[31] and Alexandre-Théophile Vandermonde.[32] Monge also gave seminars on the following topics: the manufacture of forge and case-hardened steels, arms manufacture, and the development of military balloons.

---

[29] *Ibid.*

[30] Booker, 1963, p. 23.

[31] Weller, 1999, p. 61. Monge was a close friend of Berthollet (1748–1822). Claude-Louis Berthollet was one of France's most distinguished chemists (after Lavoisier). The many credits to his name include identifying composition of ammonia (1785) and hydrogen sulfide (1789). He researched pigments and dyes and he helped to make universal Lavoisier's system of chemical nomenclature.

[32] Vandermonde (1735–1796) was a scientist, mathematician, and a violinist. He also conducted chemistry experiments with Bézout and Lavoisier.

Visiting the French factories and foundries enabled Monge to become keenly aware of just how far France was behind Britain. Monge recognized that technological progress in France would breed power and superiority as a nation. And, as a teacher, he had the foresight to know that serious changes were necessitated in the area of education. For the future of France, it was imperative to give students a robust technical education.

The practical knowledge Monge gained at the École Royale de Génie at Mézières and in his role of overseeing the naval academies guided him in the regeneration of the educational system in France. He proposed and enacted reforms to technical and theoretical instruction pertinent to manufacturing, infrastructure, engineering and the sciences. He was a true leader in this endeavor. In 1793, Monge submitted a plan to the government for the establishment of trade schools for craftsmen and tradesmen. Although his proposal was denied, he was even more determined to create a national school of higher education for training military and civilian engineers.

The Committee on Public Safety, to which Monge, Lagrange, Laplace were appointed in 1794, was charged with setting up and structuring the curriculum for the École Centrale des Travaux Publics in Paris. Monge became a prominent member of the school's governing council and the supervisor of the division of the school which trained *chefs de brigade*, or foremen. By this time, the restriction on Monge's descriptive geometry was lifted which meant he could make technical drawing the key element of such an education. Monge was an expert and passionate instructor of descriptive geometry and taught "revolutionary courses" on the subject.[33]

Since military and naval schools were not typically located in the cities, the mission of the school, for practical purposes, was ad-

---

[33] *Dictionary of Scientific Biography*, Vol. IX, 1974, p. 473. Also See Appendix B.

justed to make it an engineering preparatory school. In 1795, the school was reorganized and renamed as the École Polytechnique. For those admitted, the two-year school provided access to high level military and technical government careers. Engineering students would require two additional years of academic work in their field of specialty at one of the schools in France. A new educational system for scientific and technical training was born at the school. Monge's service, both as administrator and teacher, was an invaluable asset to all stages of the school's development. He was extremely successful at both and the true motivating force of the institution.

The École Polytechnique's strength, in addition to its ambitious curriculum, was its stellar faculty. The presence of Monge, Lagrange, and Legendre on the faculty dictated a heavy emphasis on mathematics in the first two years. Over half of the total lesson hours were devoted to the study of analytic and descriptive geometry, mechanics and analysis. Mathematical research in an educational setting was launched as an established scientific career in France at the École Polytechnique. Under Monge's directorship, the curriculum at the school became the vehicle for the current research interests of its prestigious staff of professors. The directors of the school were convinced that a solid grounding in mathematics was the best preparation for any scientific career, while continuing to value the school's original commitment to engineering. The mathematical content of the École Polytechnique's curriculum, while highly advanced, was selected and presented with an eye toward real life applications. This reinforced the school's mission to develop well-educated citizens and to develop the talent necessary to advance science in general, moulding well-rounded future leaders. A listing of the subjects and teachers present at the École Polytechnique in the first decades of the nineteenth century is provided in Appendix B.

During the spring of 1795, while Monge was setting up the curriculum for the École Polytechnique, he was also a professor at the École Normale. It was at both of these schools where Monge launched his own work in analytical, infinitesimal, and descriptive geometry. The École Normale[34] was formed in Paris in 1794 as a training school to prepare teachers. Monge refined the descriptive geometry course that he had methodically constructed at Mézières and gave his first official lectures in the subject at the École Normale. His trailblazing work became the impetus for the development of modern geometry. Monge's lectures were recorded in the *Journal des Séances des École Normales* and finally published in book form as *Géométrie Descriptive* in 1799 (see Figure 2.2).[35]

While at the École Polytechnique, Monge lectured on two mathematical subjects that were new in a university curriculum. First, he presented an intense course in his descriptive geometry[36] to four hundred students. Included in this course was the study of shadows, perspective, and topography. A significant number of his lectures addressed properties of surfaces, normal lines, tangent planes, and the theory of machines. One of the problems investigated by Monge was to construct the curve of intersection of two surfaces of revolution in which the axis of one surface is vertical and the axis

---

[34] Boyer, 1960, p. 24. The École Normale was hurriedly opened and, as a result, the school was short-lived due to administrative troubles.

[35] *Dictionary of Scientific Biography,* Vol. IX, p. 474. A friend and former student, Jean Nicolas Pierre Hachette (1769–1834) had preserved Monge's École Normale lectures in descriptive geometry and had them published in 1799 under the title *Géométrie Descriptive*. While this textbook was written for Monge's students, it became extremely successful throughout France and abroad. There were several editions; in 1811 Hachette provided additional supplements; in 1820, after Monge's death, Barnabé Brisson included new, unpublished material on Monge's theory of shadows and perspective; and a 5th edition in 1827.

[36] See Appendix B.

of the other surface intersects the first. Monge used what he called
"the cutting spheres method" for finding this curve.[37]

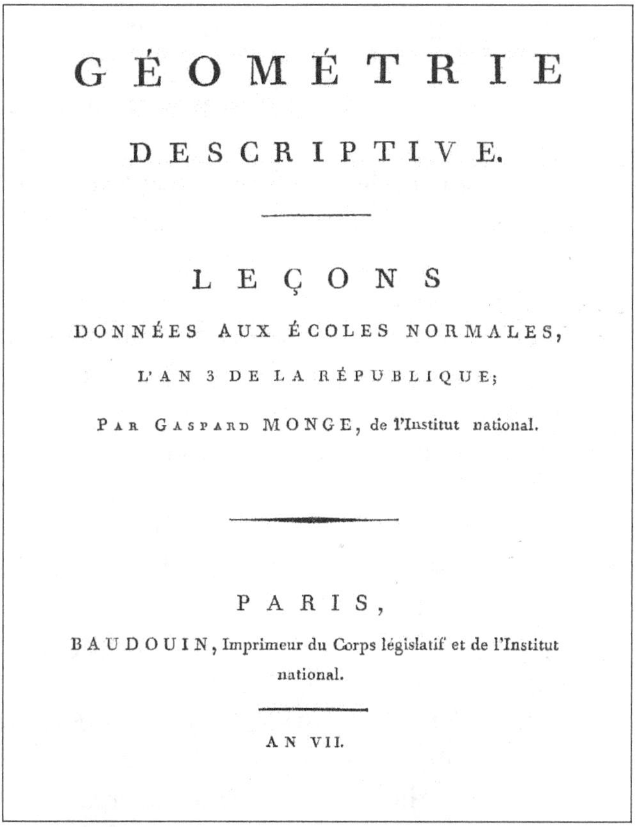

> # GÉOMÉTRIE
>
> ## DESCRIPTIVE.
>
> ### LEÇONS
>
> DONNÉES AUX ÉCOLES NORMALES,
>
> L'AN 3 DE LA RÉPUBLIQUE;
>
> Par Gaspard MONGE, de l'Institut national.
>
> PARIS,
>
> BAUDOUIN, Imprimeur du Corps législatif et de l'Institut national.
>
> AN VII.

Figure 2.2: *Géométrie Descriptive* (1799)

Descriptive geometry was by no means Monge's only contribution to three-dimensional geometry. The second subject Monge taught at the school was a course in the application of analysis to geometry. Essentially, it was an introductory course in differential geometry of curves and surfaces in three dimensions. It can be

---

[37]Monge, *Geometrie Desctriptive*, 1799, Chapter III.

asserted that the seventeenth century was the century of curves: the cycloid, the limaçon, the catenary, the lemniscate, hyperbolas, parabolas, and a copious number of other new curves, while the eighteenth was the century that really embarked upon the study of surfaces.[38]

Since there had not been a textbook written on the application of analysis to geometry, Monge, who did not relish the idea of writing textbooks, compiled his lecture notes which he called *Feuilles d'analyse appliquée á la géométrie* (1795) for students to use in the classroom. He knew that creating suitable and useful textbooks was vital to reforming the curriculum at the École Polytechnique. Specifically, textbooks containing his analyses and applications were in great need as the course became a requirement for all students at the school. In the class on applications of analysis to geometry, Monge brought solid analytic geometry into the forefront. His lectures encompassed a brief presentation of the fundamentals of lines and planes, with the major portion of material taught on calculus and its applications to curves and surfaces in three dimensions. In this area of mathematics, Monge was particularly interested in developable surfaces and was the first to distinguish between a ruled and a developable surface. He showed that a developable surface is one that is the envelope of a one-parameter family of planes. Given a parameterized family of curves, the boundary of the region swept out by the curves in the set creates a new curve called an envelope. An envelope of a family of curves is tangent to each curve at a point. Visually speaking, the surface can be "unrolled" onto a flat plane. Conversely, a developable surface can be formed by bending plane regions. For example, Figure 2.3 illustrates the development of a prism (cylinder) and a pyramid (cone).

---

[38]Boyer, 1960, p. 23.

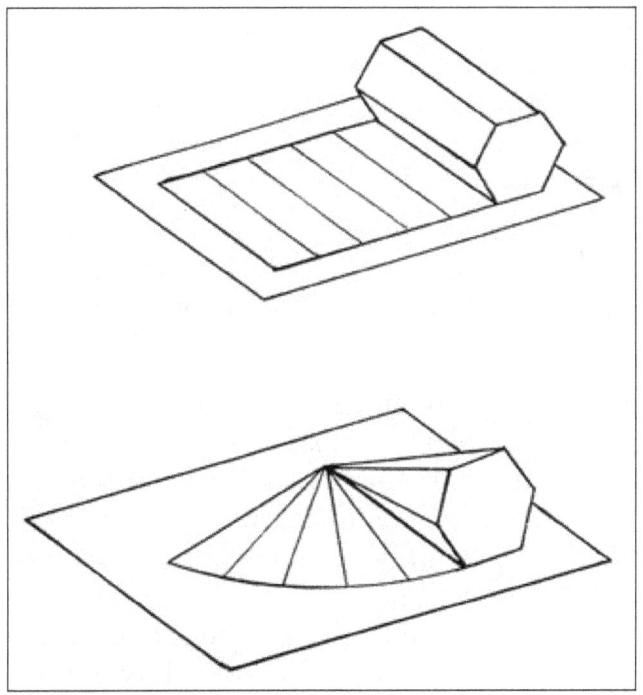

Figure 2.3: Cylinder and cone as developable surfaces

The character of the École Polytechnique was immeasurably influenced by Monge in its merging of the theoretical sciences with manufacturing arts. The instruction was (and still is, today) based on a well-ordered and rigorously demanding curriculum. The school hired the most outstanding mathematicians of the period. In the classrooms and laboratories, the young students were exposed to the personal influence of important teachers. Thus, the fact that the achievements of the students in the school were extraordinary was not surprising. Felix Klein noted:

> The external influence of the school was increased by a rule that lectures had to be published. The great majority of the leading textbooks in higher mathematics at the beginning of the 19th century arose from the

instructional activity a the Ecole Polytechnique. Included in the curriculum were mechanics and mathematical physics, geometry, analysis, and algebra. Such intensive activity could not fail to have an influence on all of science.

In fact, almost everything achieved in mathematics, physics, and chemistry in France during the first decades of the 19th century had its origin in the Ecole Polytechnique.[39]

The students received supplemental guidance from *répétiteurs* who would further explain the material and quiz students on the lectures given by the professors. Every student took a cumulative and comprehensive final examination written by the *examinateurs*. Their task was to ascertain what the students had mastered. The École Polytechnique became the model for colleges of engineering throughout Europe and also for the United States Military Academy at West Point. Monge's sense of duty and intense interest in science was consonant with his eminent pedagogical qualities. The École Polytechnique would not have thrived without Monge's uplifting style of teaching, his dedication to the profession, and his political activities.

2.6 Monge the Professor

The original mathematics Monge produced and published in geometrical research became a vehicle of study for his students and followers. The influence of his teaching proved equally fruitful. Monge knew how to present the material with precision and passion. His students were left with enthusiastic memories of his geometry lessons. Joseph Fourier (1768–1830), a student of his at the École Normale, wrote in a letter describing Monge as his teacher:

---

[39]Klein, 1979, pp. 61–62.

> Monge has a strong voice, he is active, ingenious, and very knowledgeable. He excels in geometry, physics, and chemistry; the sciences in which he gives his lessons with most clarity and of which he is infinitely curious. At the same time, one finds that he is too clear or rather that his method (of explanation) is not rapid enough. He gives particularly practical lessons. He speaks familiarly (of his subject), with precision more often than not. He is not only known for his advanced knowledge, he also speaks very easily about public and private matters. His outward appearance is quite ordinary.[40]

As Fourier commented, Monge, at first sight, emerges as a rather unremarkable looking professor. However, when in the midst of teaching, his livelier traits would appear. His visage would become immediately illuminated, replacing his normally stoic countenance, with animated expressions as his eyes glowed with intensity. When Monge described a surface, he generated it with his hands in an eloquent gesture that rendered it tangible for his students. The abstractions of the mathematics took shape as he made the most complicated concepts seem simple and the most obscure theories more clear. Monge was able to do this, because he knew the logical relationship of the ideas he was presenting from the most elementary to the most profound.

In the lecture hall, Monge could discern rather quickly if some students hesitated to understand the material and he would begin again with a different form of reasoning. His essential goal was clarity and he achieved it, and rediscovered it, each time he presented a lecture. At the end of a course, Monge did not consider his mission to be over. He would continue his discussions with the less confident students to try and resolve their last difficulties, and guided the leading research of his most talented students. He was

---
[40]Taton, 1951, pp. 366-67.

encouraging to all with his advice and worked hard at teaching effectively to develop each student's ability to undertake new research along the road that he himself had built. This inveterate teacher inspired his students to greatness.

François Arago (1786–1853) was a student whose admiration for Monge led him to write a biography of the famed geometer. Arago knew Monge quite well and loved him equally for his "la fougue et la douceur," bravado and kindliness. Arago viewed Monge as:

> ... a teacher who, endowed with special skills, imparted techniques he himself had found and taught them to his students as a set of discoveries both determinate and applicable. He would stand be fore them, a hero of their own world, indeed a maker of it, his achievements a rebuke of what they too lacked—grace, deportment, ease, in a word a liberal education. Ever a teacher, Monge in return prized that which he thought to have instilled in younger men—talent, skill, address, ambition. He was one of those teachers who side with the rising generation against their own, and who reassure the new world by joining in subversion of the manners of the old. Enveloped in his concern for youth, immaturity merged into vigor, rebellion into pride, vulgarity into a concern for humanity.[41]

---

[41]Gillispie, 2004, p. 341.

2.7 French Politics and Monge's Mathematical Career (1796–1818)

Monge's political life played an unusual and decisive role in the development of his mathematics.[42] When the Bastille fell on July 14 in 1789, Monge was an eager supporter of the Revolution and he joined in its spirit by attempting to recruit new patriots for the future regime. Monge was a leader in the scientific uprising as well as in the revolt against the government. He embarked on a surprising friendship with Napoleon Bonaparte in 1796 when he was named a member of the Commission des Sciences et des Arts en Italie. Monge and other associates in the Commission (of which Berthollet was also a member) were sent on a mission to Italy to complete the distasteful task of bringing prized paintings, manuscripts, and sculptures back to France. He traveled to northern and central cities of Italy, including Rome. During his visit, Monge became experienced in cheese making processes from local farmers. The *Annals of Chemistry* (1799) contains a detailed essay Monge wrote on making homemade Parmesan cheese entitled, "Sur la fabrication du fromage de Lodézan, connu sous le nom de Parmézan." While on assignment in Italy, he was heavily encouraged to take a treasured work of art for himself. But Monge could not be bribed. He showed himself to be a public servant above reproach. He left Rome with a clear conscience and brought back to France the priceless objects (including the Mona Lisa). Monge and Napoleon established a close kinship during and after their travels

---

[42]Smith, 1932, pp. 111–122. This article provides an account of the political life of Monge based on first-hand information given in eighty unpublished documents relating to Monge and his family. Half of the letters were written by Monge himself. The remaining letters were written by his brother Louis or his wife. These letters, along with portraits and medals of Monge, reside in the Columbia University Rare Book & Manuscript Microfilm #80-1582.

together and they maintained frequent correspondence with each other. Napoleon was enthusiastic about mathematics and showed aptitude in it since his teenage years. His interest in mathematics and applied science grew even stronger over the years. This may make somewhat clear the affinity Napoleon had towards Monge and other scientists.

The following year, in 1797, Monge returned to Paris for a brief stay to undertake a new appointment as the director of the École Polytechnique. In 1798, Monge, other scholars who were confidants of Napoleon, and a group of his own students from the École Polytechnique were requested by Napoleon to accompany him on an expedition to Egypt. The French fleet of five hundred ships descended upon Malta in June of that year and captured the city. Napoleon's aim was to civilize Egypt, that is, "to bestow on its people all the benefits of European civilization."[43] Napoleon and Monge achieved this goal without procrastination. As an initial step, in one week Monge set up fifteen elementary schools and a school for higher education, the Institute d'Egypte, rooted in the spirit of the École Polytechnique. In addition, Monge's students from the École Polytechnique, under his, Fourier's, and Berthollet's direction, performed a geodesic survey of the country.

Later that year in Cairo, Monge was appointed President of the newly established Institute d'Égypt. He traveled with Napoleon on various other (some dangerous) expeditions in Egypt. Militarily, things began to go awry for Napoleon. At the end of 1799, he made an unexpected and covert return to France with Monge and Berthollet, who were later cited for their bravery. Monge was reinstated in his position as head of the École Polytechnique. It was disclosed, through various communications and correspondence during Monge's time away at the Institute d'Egypt, that he had been perfecting the chapters of *Feuilles d'analyse appliquée á la géométrie*.

---

[43]Bell, 1945, p. 193.

Upon his latest return to the École Polytechnique, Monge found the school to be severely financially burdened. The zeal of the previous government disappeared with the new Emperor of France, Napoleon. In order to keep the school functioning, Monge contributed his salary and even his retirement annuity to support those students who could not afford the tuition. Napoleon swiftly had a change of heart after becoming aware of Monge's actions and he helped the school out of its financial distress. Then, Monge, the foremost leader of the new geometry, became a preeminent supporter of the imperial regime. In the years following the Egyptian campaign, Napoleon bestowed many honors and appointments on the scientists who assisted him. He honored Monge with the title of Compte de Péluse (1804), appointed him as Grand Officer of the Légion d'honneur (1804), and nominated him as President of the Senate in Liége (1806). Napoleon ensured that Monge was self-supported, he presented Monge with an estate in Westphalia, and, on his retirement, paid him two hundred thousand francs.[44]

Over the next few years, Monge continued teaching, research, and writing textbooks on the applications of geometry to calculus for the students at the École Polytechnique. In 1807, Monge completed the final version of his *Feuilles d'analyse appliquée á la géométrie* and formally published it as *Application de l'Analyse á la Géométrie*.[45] To Napoleon, the mathematical, scientific, and technological work of Monge and his fellow scientists enriched the lives of others and improved the economic status of France as a nation. Napoleon was drawn to scholarly individuals who could be of use to him in solving practical problems. In return for their services and loyalty, he greatly rewarded these prominent scientists in their personal lives.

---

[44]Booker, 1963, p. 29. Also, Smith, 1932, p. 120.

[45]*Dictionary of Scientific Biography,* Vol. IX, p. 474. This work was republished in 1809 and its last edition was published 1850 with an introduction and appended important supplements by Joseph Liouville (1809–1882).

In 1812, Napoleon's success was beginning to wane. The Austrian and Russian military offensives were foiled and costly, both in terms of lives and money. When Napoleon was defeated, so followed his loyal supporter, Monge. Monge's honors and appointments were all taken away. When Monge died during the summer of 1818, the political defeat he experienced was transcended by his mathematical brilliance and the esteem and appreciation felt towards him by his scientific contemporaries and students.

Chapter 3
Monge's Mathematics

> Monge was one of the first modern mathematicians whom we recognize as a specialist: a geometer—even his treatment of partial differential equations has a distinctly geometrical touch. Through Monge's influence geometry began to flourish at the École Polytechnique. In Monge's descriptive geometry lay the nucleus of projective geometry, and his mastery of algebraic and analytical methods in their application to curves and surfaces contributed greatly to analytical and differential geometry.[1]
>
> Dirk Struik

3.1 Descriptive Geometry

Gaspard Monge's mathematical acclaim arose from his reinvention of the subject of descriptive geometry and his creation of the subject of differential geometry of curves and surfaces. Monge's work in the "geometrization of mathematics" focused on two central ideas: i) the correlation of analytic operations with geometric

---
[1] Struik, 1967, p. 147

transformations in his axioms of descriptive geometry and ii) the generation, parameterization, and classification of surfaces which gave rise to the subject of differential geometry.[2] It was primarily through Monge's own teachings that descriptive geometry, analytical geometry, and differential geometry were established as special fields.

Monge established descriptive geometry as a key subject in technical education and perceived that machines were impossible to build unless they had truly been designed on paper.[3] It was considered to be a foundational subject used to train prospective draftsmen, architects, engineers, and also mathematicians. This graphical and mathematical language that Monge created became an accepted method of communication between engineers and designers.[4] The five core chapters in *Géométrie Descriptive* (1799) are comprised of topics on: the objective and principles of descriptive geometry and problems relative to lines and planes (Chapter I), tangent planes and normal planes to surfaces (Chapter II), intersections of surfaces and definitions of curves of double curvature, i.e., space curves (Chapter III), applications of intersections of surfaces to the solution of diverse problems (Chapter IV), plane curves, space curves, and surfaces (Chapter V).

3.1.1 The Mongean Method

In Chapter I of his textbook, Monge outlines the two main of objectives of descriptive geometry:

> First, to give an understanding of the methods of representing on a two-dimensional surface objects which in

---

[2]Glas, 1986, p. 257.

[3]See Appendix C for a translation of the *Programme* to *Géométrie Descriptive* (1799) in which Monge states his general plan for the necessity of teaching the subject.

[4]Cardone, 2003, p. 68.

nature have three dimensions... the second objective is to teach a way to determine the forms of objects and to deduce all of the properties resulting from their respective positions."[5]

To achieve these objectives, Monge built his work around two complementary geometric conventions. The first essential principle centered on horizontal and vertical projections. Monge perceived that any point in space could be defined by three coordinates determined by the distance of the point to three mutually perpendicular planes. Taking into consideration that each coordinate of the point could be determined by its projection into two of these planes, only two planes were really necessary. Thus, throughout his constructions Monge used the horizontal and vertical projections of the point in plan and elevation. In order to make the renderings even more straightforward for draftsmen to complete, Monge revolved the vertical plane into the horizontal plane so that both planes would be shown on one sheet of paper, which gave two side-by-side complementary views. These double orthographic projections allowed him to completely and uniquely describe a three-dimensional object in the plane. Figure 3.1 illustrates these basic principles of descriptive geometry.

---

[5]Gafney, 1965, p. 556.

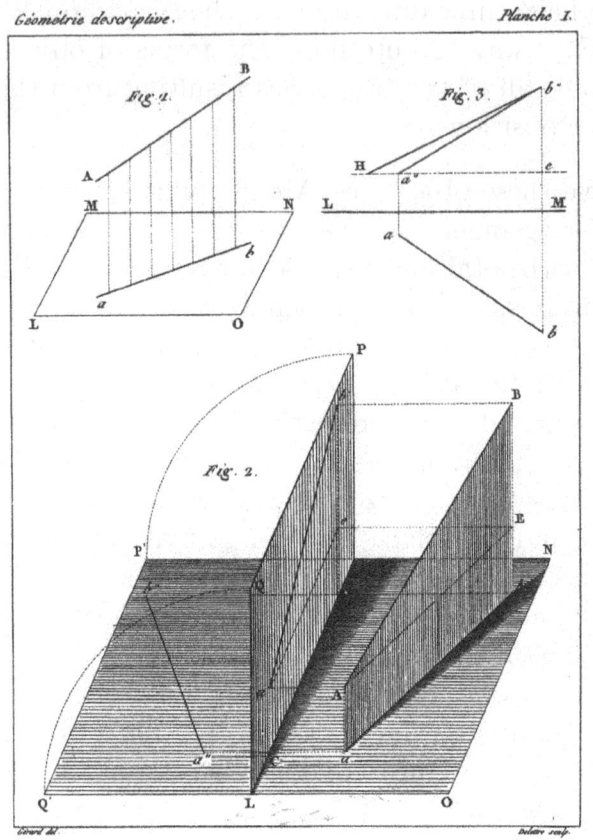

Figure 3.1: Monge's construction of double orthographic projection from *Géométrie Descriptive* (1799)

This new method used exact geometrical constructions and replaced the previously demonstrated methods which were based on inexact perspective sketches. Double orthographic projection techniques date back to the work of famous German painter Albrecht Dürer (1471–1528) and a brilliant French military engineer, Amédée-François Frézier (1682–1773) whose specific interests centered on curves of intersection of surfaces. Although this approach was not new to other experienced mathematicians, it was Monge who developed what was initially a set of "rules of thumb" into a universal system and exact mathematical procedure.[6]

A second major principle of Monge's descriptive geometry was the generation of surfaces. In pre-Mongean drawings, a cone would be represented by a triangle in elevation (side view) and by a circle in plan (top view). In terms of Mongean geometry, the two sides of the triangle were the generators of the surface being revolved about the center line, or the axis of generation, and the circle in plan was the trace (curve of intersection) of the surface generators on the horizontal plane. Alternatively, the cone could be generated in another way by considering the circle in plan as the generator whose radius diminished in moving up the cone. In other words, the diameter of the circle would fit into the triangle at any height. Monge's concept of generation viewed solid geometrical shapes as the result of movements in space—not as static figures. It was on this premise that the method Monge used for determining the curves of intersection of two surfaces was based.

3.1.2 Examples from *Géométrie Descriptive*

In *Géométrie Descriptive*, Monge conscientiously provided a variety of well thought out problems with solutions. Each stated problem was designed as a model for students to replicate and acquire expertise in solving. Monge's techniques used the simplest and most economical constructions possible in order to minimize the work required by the draftsman, and, at the same time, to ensure maximum exactness. He wrote in his textbook:

> Indeed, since it is well known that every line drawn inevitably introduces a source of error, it follows that a reduction of the number of lines that must be drawn

---

[6]Cardone, 2003, pp. 67–68. Sylvestre Francois Lacroix (1765-1843), a well-known student of Monge, published a book on descriptive geometry in 1795.

in a given construction increases the accuracy of the construction.[7]

A common problem Monge solved was to determine the projections of intersections of curved surfaces. In the design of machines and other objects, drawings showing the lines of intersection of the various parts were required in order to represent the object in its space. Monge labeled the intersection of two surfaces as *une courbe á double courbure* or a space curve (i.e., a curve in three-dimensional space). Fifty figures were included in the book to elucidate the system he employed in solving various problems. Types of problems he chose were: the intersection of two cones whose axes are parallel, the intersection of two surfaces of revolution whose axes are concurrent, the intersection of two cylinders, and the intersection of two cones or of a cone and a sphere with a tangent plane.

Monge made the connection that exists between the operations of analysis and the methods of descriptive geometry. He addressed the parallelism between the geometric determination of finding the curve of intersection of two surfaces and the concept of algebraic elimination.[8] He asserted that it was upon this correspondence that the foundation of his principles of descriptive geometry was constructed which later led to his transition to the study of surface theory. As Monge stated:

> In order to learn mathematics in the most advantageous manner, it is necessary that the students early on become accustomed to becoming aware of the connection that exists between the operations of analysis and those of geometry; it is also necessary that the students be

---
[7]Roever, 1933. p. vi.
[8]Monge, *Géométrie Descriptive* 1799, pp. 61–62.

able to write in analysis all the movements that can be conceived of in space, and, conversely, to perpetually visualize in space the geometrical representation for which each of the analytical operations is written.[9]

The next two sections describe in detail two of Monge's geometric construction methods, the cutting planes method and the cutting spheres method.

Cutting Planes Method

The first problem considered is to construct the projections of the curve created by two intersecting surfaces of revolution. Monge first takes on a general approach, then modifies the procedure, accordingly, to deal with specific surfaces such as cones or cylinders. He assumes that the shape of each surface is known since he is able to construct their generators through any given point. Monge gives the following general statement of the problem:

> We return to our objective, which is the method of determining the projections of intersections of curved surfaces. The wording will be general and applicable to any two surfaces; and although the letters that we use will relate to Figure 26,[10] which represents the particular case of two conic surfaces with circular bases and vertical axes, it is nonetheless necessary to always conceive that each particular surface may be one other than a conical surface.
>
> *The first general problem.* Suppose that the [method of] generation of the two curved surfaces is known, and

---

[9]*Ibid.,* pp. 61–62.
[10]See Figure 3.2.

> given that all of the [information] which fixes these generated surfaces is determined on the planes of projection, construct the projections of the curve of double curvature [the curve of intersection is a space curve as we know it] which intersects the two surfaces.[11]

In the explanation of his solution, Monge continues:

> In order to construct the intersection of two surfaces of revolution in which the axes are vertical, the most advantageous method uses a system of cutting planes consisting of a sequence of horizontal planes, because each plane cuts the two surfaces in circles of which the centers are on the respective axes, and of which the horizontal projections are the circles known by size and position, based on the height of each cutting plane. In this case, all the points in the horizontal projection of the intersection of the two surfaces are found by the intersections of arcs of circles. One notes that if the surfaces of revolution have their axes parallel to each other, but not vertical, it will be necessary to change the planes of projection and to choose them in a manner such that one of them will be perpendicular to the axes.

> Generally, in solving most problems involving the intersection of surfaces, a series of cutting planes is used which cut either straight-line or circular elements on the surface. These elements lie on the same cutting plane as well as on their respective surfaces and intersect to create points on the curve of intersection of the two surfaces. If the series of cutting planes are spaced properly apart, a sufficient number of points will be constructed through which the line of intersection may be drawn.[12]

---
[11]Monge, *Géométrie Descriptive* 1799, p. 62.
[12]*Ibid.*, p. 64.

In his solution, Monge first sets up the construction for the two conical shaped surfaces under consideration as shown in Figure 3.2. The upper half of the page shows a side view (in elevation) of the two intersecting cones, their respective axes of revolution, and the series of cutting planes labeled EE', ee'. The lower half of the drawing shows the top view (in plan) of the two cones represented by circles XRZ and UVT. In this particular drawing, Monge assumes that the cone on the right-hand side (with vertex b in elevation) is slightly in front of the cone on the left-hand side (with vertex a in elevation). This is explicitly illustrated in the top view by noting that the axis ZX, the horizontal projection of the cone with vertex b (or circle XRZ with center B), is below the axis TU of the horizontal projection of the cone with vertex a (or circle UVT with center A). To sketch the projections of the curve of intersection for two surfaces of revolution whose axes are parallel, Monge first conceives of a series of horizontal planes, perpendicular to the axes of revolution. The vertical projections (in elevation) of these planes is a series of horizonal lines identified as EE', ee'.

We follow Monge's construction using the particular cutting plane labeled EE'. Plane EE' intersects the first surface in a certain curve, which Monge states will always be possible to construct, since this curve is formed by the series of points in which the plane EE' is cut by the generatrix (or ruling) in all of its positions. The horizontal projection (in plan) of this curve is given by the circle $FGHIK$ in Figure 3.2.

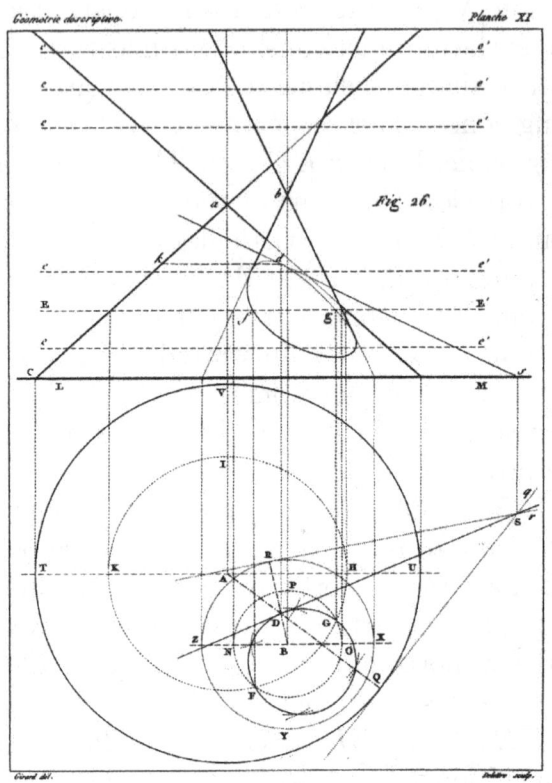

Figure 3.2: Monge's cutting planes method for determining the curve of intersection of two cones from *Géométrie Descriptive* (1799)

The same plane EE' cuts the second surface in another curve, and its horizontal projection will be the circle $FOGPN$. Once these constructions are made, either the projections of the two curves, given by the circles $FGHIK$ and $FOGPN$ intersect, or they do not intersect. If they do not intersect, then at the height of the plane EE', the two surfaces do not have any common points. However, if the two curves intersect, they will do so in a certain number of points which will be common to the two surfaces. In Figure 3.2, we see that the points F and G, which meet the two circles $FGHIK$ and $FOGPN$, are the horizontal projections of two points of intersection of the two surfaces of revolution. In order to find the vertical projections of the same points, since they are contained in the plane EE', then their projections must be on the straight line EE'. So, the points F and G can be orthogonally projected onto EE' in the points f and g, which are the vertical projections of the two points of intersection.

Monge follows the same process (as was done EE') for all the other horizontal lines labeled ee'. The result is a series of new points $F\ G,\ldots$ in the horizontal projection and a series of new points f, g, $\ldots$ in the vertical projection. The trace of the curve through the new points F, $G,\ldots$ is the circle $FOGD$ in plan. Similarly, in elevation, Monge drew a smooth curve through the new points f, $g,\ldots$ which delineated the vertical projection of the curve of intersection, indicated by curve fgd in Figure 3.2, that was desired. The process Monge applied for finding a sufficient number of points, all lying on the projections of the curve of intersection, was dependent upon the construction of an adequate number of cutting planes. In his final step, Monge drew a smooth curve through all of those points to obtain the curve of intersection.

Another problem Monge considers, using the cutting planes method is the construction of a conic section, that is, the curve of intersection of a cone and a plane.[13] In Figure 3.3, Monge determines an ellipse and its horizontal projection.

---

[13]Monge, *Géométrie Descriptive* 1799, p. 73, No. 69, *Deuxiéme question*.

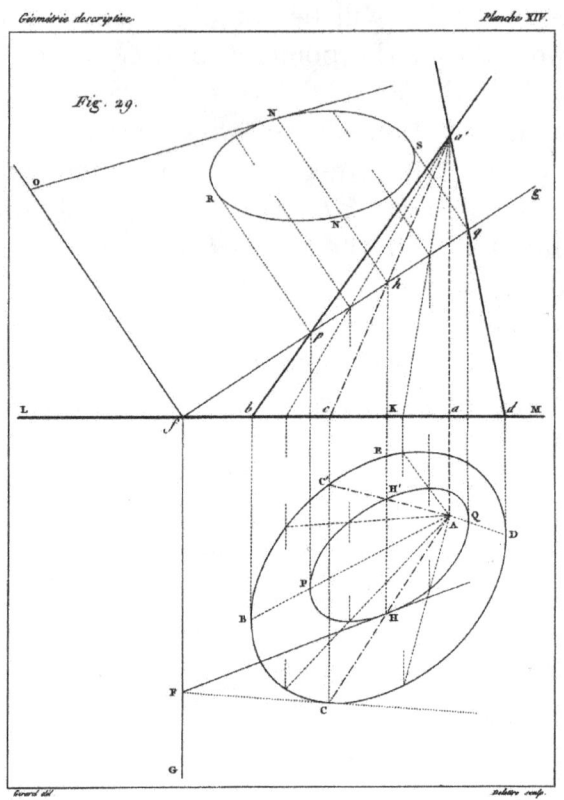

Figure 3.3: Monge's cutting planes method illustrating the construction of a conic section from *Géométrie Descriptive* (1799)

Cutting Spheres Method

Monge showed that, in certain cases, choosing the proper system of planes or even using a sequence of curved surfaces, differing from the system of planes by one dimension, plays a significant role in producing the most simple and elegant constructions. In Figure 3.4, Monge used a series of concentric cutting spheres to find the intersection of two surfaces of revolution whose axes intersect. The point of intersection of the two axes was used as the center of the series of cutting spheres. Each sphere cuts circles on each surface of revolution. If both axes are at right angles in the line of sight, then the circles will appear in their edge view on the cutting spheres. That is, the vertical projection of each sphere will be an arc of a circle that is perpendicular to each of the axes. The points on the curve of intersection of the two surfaces are where the edge views of the circles intersect. In Figure 3.4 (Monge's Figure 35), the two surfaces of revolution Monge generated were projections of ellipsoids. Although Monge referred to specific surfaces in his drawings, the process he used was general and systematized so as to be used for any two types of intersecting surfaces of revolution.

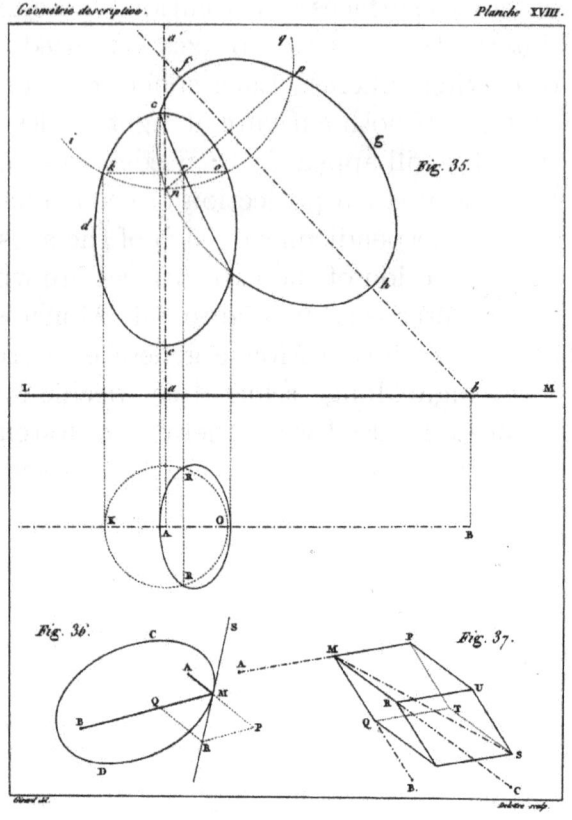

Figure 3.4: Monge's cutting spheres method for determining the curve of intersection of two surfaces from *Géométrie Descriptive* (1799)

Monge sets up his construction in Figure 3.4 in the following manner. The side view representations of the two intersecting surfaces of revolution and their respective axes of revolution intersecting at a point are shown in the upper half of the drawing. Surfaces are arranged to make the axis of revolution of one surface, in this case the surface generated by the arc *cde*, perpendicular to the horizontal plane of projection. In the lower half of the drawing, which represents the horizontal projections, other pertinent information relating to the set up of the construction diagram is given: point A is the horizontal projection of the axis of the surface generated by cde, $a'a$ is the vertical projection of this axis; the line AB is parallel to LM; the horizontal projection of the axis of the surface generated by fgh and $a'b$ is the vertical projection of its axis of revolution; and A and a' are the projections of the point of intersection of the two axes. Next, Monge constructs a series of concentric cutting spheres whose common center is placed at the point of intersection of the two axes of revolution. For each cutting sphere, Monge shows its partial vertical projection as an arc of a circle in side view (elevation). One such projection is given by the arc of the circle $iknopq$.

Each cutting sphere will intersect the first surface of revolution, the surface generated by cde, in a circle shown in edge view labeled as kro. The edge view of this circle will be perpendicular to the axis $a'a$. The vertical projection of circle kro is given by the segment KO in the top view (plan) drawing. The point A is the midpoint of the segment KO, and the center of the circle $KROR'$ with diameter KO. Note that the circle $KROR'$ is the horizontal projection of the circle kro given in edge view in the elevation drawing. Similarly, each cutting sphere will intersect the second surface of revolution, the surface generated by fgh, in a circle whose edge view is labeled nrp.

Once this process is completed for one cutting sphere, the two circles, kro and nrp will intersect in two points. The two points of intersection are represented by the point r which is their common vertical projection. Upon repeating these constructions for several cutting spheres, the curve taken through all the points r will be the vertical projection of the intersection of the two surfaces. Projecting the point r onto the circle $KROR'$ in the points R and R' will give the horizontal projections. The curve taken through all points

R, R' will be the horizontal projection of the curve of intersection of the two surfaces.

3.1.3 Monge, Théodore Olivier, and String Models

Monge was an enterprising teacher in his descriptive geometry courses. He devised a way to make the transition from lecturing with drawings to demonstrating with physical models. His goal was to help students attain a clear and vivid intuition of the surfaces they were studying. To make the geometrical discoveries real for his students, Monge built three-dimensional models from silk thread stretched over curved frames to display surfaces of revolution. Students were able to see and physically hold a doubly ruled hyperboloid of revolution of one sheet and a doubly ruled hyperbolic paraboloid. His models were still in existence at the École Polytechnique in 1814.[14]

---

[14] *Catalogue de Collections du Conservatoire des Arts et Métiers*, 1851 p. 17.

Figure 3.5: Olivier string model of a ruled cylinder at the Musée des Arts et Métiers

The surface representations Monge made were further advanced by his student Théodore Olivier (1793–1853). Olivier was taught by Monge at the École Polytechnique. In 1830, Olivier devised and fabricated over fifty interactive models, all with moving components, to show ruled surfaces of revolution and their intersections. He was undoubtedly influenced by his mentor, Monge, and his own creations in descriptive geometry.[15] Olivier believed in Monge's philosophy that students needed to visualize and manipulate the geometric surfaces they were studying in order to truly understand them. These physical figures were the pedagogical tools designed exactly for that purpose. Whereas Monge's models were fixed and static, Olivier's models were movable and dynamic. They could be rotated about one or more axes to show a wide range of geometrical configurations and intersections of surfaces.[16]

---

[15]Stone, 1969, p.2. In addition to being noteworthy teaching aids, mathematicians and artists find the models to be beautiful works of art.

[16]*Ibid.*, p. 2.

Olivier was one of the founders of the École Centrale des Arts et Manufactures in 1830. He later became professor of descriptive geometry at the Conservatoire National des Art et Métiers in 1839 where he lectured on perspective, cross-section, and the intersection of two surfaces using the string models he designed. Olivier wrote a number of works on the subject in descriptive geometry including *Développments de Géométrie Descriptive* (1843), *Mémoires de Géométrie Descriptive Théorique et Appliquée* (1851), and *Cours de Géométrie Descriptive* (1852–1853). A Paris firm was commissioned by Olivier to construct two complete sets of his models. Several of the original Olivier models are still in existence at the Musée des Arts et Métiers in Paris and in the United States at Union College in Schenectady, New York.[17] The Department of Mathematics at the United States Military Academy at West Point had copies made, by the same Paris firm, and currently maintains twenty-four models. There are also copies of the Olivier models housed in the Harvard University Collection of Historical Instruments. Other models are stored at the Canada Science and Technology Museum and in the London Science Museum. In designing his string models, Olivier did what every mathematics professor desires for his students: he surpassed his teacher!

3.1.4 Descriptive Geometry After Monge

Historians of mathematics and science deem Monge's lessons at the École Normale to be birth of descriptive geometry. The far-ranging success of Monge's *Géométrie Descriptive* launched a new category of geometry throughout France and overseas. Although, some

---

[17]Stone, 1983, p.11. The Olivier models at Union College are an original set. They were purchased in 1856 from Madame Olivier by William Gillespie, a Professor of Civil Engineering (from 1845–1868), who personally knew Olivier. The college formally procured the models in 1868 and were used in classes, as intended by their inventor, until the end of the nineteenth century.

historians[18] considered the subject closed after Monge's former student Jean Nicolas Pierre Hachette (1769–1834)[19] published a set of supplements to the book in 1811 and 1812. Hachette regarded the addenda to be the final touches to Monge's original astonishing work.

Monge considered descriptive geometry as a tool for developing his original ideas in pure, infinitesimal, and analytical geometry. From 1771 to 1809, Monge produced many papers on infinitesimal geometry in three dimensions. His work in analytical geometry began more definitively in 1795 with his lectures at the École Polytechnique. In 1802, Monge and Hachette published "Application de l'algèbra à la géométrie" in Volume 4 of the *Journal de l'École Polytechnique*. Problems on change of coordinates, the theory of conics and quadrics, and a summary of Monge's lectures in solid analytic geometry were presented in this important memoir (which appeared again in 1805 and 1807). Monge published two papers in 1809 and 1811 entitled "Sur la pyramide triangulaire." Included in the results is *Monge's Theorem*:

> The planes drawn through the midpoints of a tetrahedron perpendicular to the opposite edges meet at a point M (called the *Monge point* of the tetrahedron). M turns out to be the midpoint of the segment joining the centroid and the circumcenter.[20]

---

[18] Coolidge, 1940, p.114.

[19] Hachette was a draftsman at the École Royale du Genie at Mézières in the 1780s. He assisted C. Ferry who was Monge's successor in the teaching of descriptive geometry at Mézières. Hachette was Monge's steady worker who eventually became his assistant, and then his successor in teaching descriptive geometry at the École Polytechnique from 1796 until the end of Napoleon's regime. It was Hachette who retained Monge's lectures from 1795 and had them published in 1799, at the request of Monge's wife, as *Géométrie Descriptive*

[20] Boyer, 1960, p. 25.

Another theorem from analytic geometry which carries his name is:

> The locus of the vertices of the tri-rectangular angle whose faces are tangent to a given quadric surface is a sphere (known as the *Monge sphere* or director sphere of the quadric).[21]

There are notable examples of new results from Monge's successors obtained from his theories in descriptive geometry. The concept of a ruled surface, an osculating plane, the radius of curvature, and lines of curvature were outgrowths of application problems Monge posed in Chapter IV of *Géométrie Descriptive*, "Applications des intersections des surfaces a la solution de diverses questions." Hachette, Olivier, Charles Dupin (1784–1873), Jean-Victor Poncelet (1788–1867), and Michel Chasles (1793–1880) were among those who followed Monge and furthered the study of these ideas.

In an 1816 article, Hachette presented a method for determining the osculating plane at a given point of a space curve formed by the intersection of two surfaces which intersect orthogonally.[22] Hachette stated the following in his solution:

> If point p is a point on the curve of intersection on the first surface with surface normal N, the vector N will lie in a plane tangent to the second surface at the point p. So the vector N will be the tangent vector to the second surface at the point p. The plane spanned by these two vectors is called the osculating plane.[23]

As a continuation of this theory, Hachette also computed the center of curvature. Dupin, Poncelet, and Chasles did similar work

---

[21] *Ibid.*, p. 25.
[22] Hachette, 1816, p. 25.
[23] *Ibid.*, p. 25

in the early 1800s on the same topics, but each presented different methods for determining the osculating plane and center of curvature. The determination of lines of curvature of a curved surface, which Monge spent a great deal of his time developing in 1796, provides another example of his transition from descriptive geometry and extension to surface theory. Dupin, in an 1806 article, found the radius of curvature at a point on a surface where the lines of curvature intersect.

Descriptive geometry was closely connected with the rise of projective geometry in the early nineteenth century. Monge revived interest in synthetic methods and attracted attention to the method of projection.[24] Dupin, Poncelet, Chasles, Lazare Carnot (1753–1823), and Joseph Gergonne (1771–1859) all attended Monge's lectures either at Mézières or at the École Polytechnique. They were all (with the exception of Chasles) career engineers keenly involved in using geometry to solve practical problems in mechanics, upon which descriptive geometry originated. Each of these men made relevant contributions to the development of projective geometry. Chasles published an abundance of original essays in the *Journal de l'École Polytechnique* on the general theory of homography and of duality, for instance.[25] Thus, the theoretical richness of descriptive geometry and its consequences illustrate how Monge's lessons were mathematically timeless. It is clear that without descriptive geometry the engineering sciences would not have developed as rapidly as they did.

Mongean *Descriptive Geometry* remains the unique method upon which contemporary technical and mechanical drawing is based. It succeeded in developing an individual's ability to imagine

---

[24]Daston, 1986, p. 269. Synthetic geometers used concepts akin to transformation, projection, and continuity to find geometric counterparts to analytical forms such as infinitesimals and negative numbers.

[25]Cajori, 1991, p. 292. A homography is a transformation from one projective plane to another. Duality is a transformation that maps lines and points to points and lines. The term *duality* is attributed to Gergonne.

and conceive of forms and volumes in space. Computer-aided design (CAD) and other mathematics software with graphical output have re-conceptualized Monge's descriptive geometry and continued it into the 21st century. Imaging and analyzing three-dimensional objects date back to the two main objectives of the subject and its founder, Monge.[26] The present day modeling software that designers and engineers use is highly sophisticated and requires a strong background in the basic principles of the theory in order to image and analyze, understand and synthesize, and grasp the mathematics of rendering three-dimensional objects. The existence of CAD and similar software in product and industrial design proves that Monge's work was the cornerstone of engineering drawing and that it is still useful today.[27] Remarkably, the procedures Monge utilized in his drawings and silk thread models are applicable in the teaching of Euclidean and hyperbolic geometry courses. The basic ideas of his fundamental constructions can be implemented and manipulated with dynamic geometry and algebra programs such as Geometer's Sketchpad and GeoGebra, which provide effective teaching tools in undergraduate geometry. These programs automate the process of drawing an object or curve by performing a single construction repeatedly, in rapid fashion, and with least effort. Thus, various parameters of a construction sketch become freely movable, making the drawing interactive and a great investigative device for students and teachers.

3.2 Differential Geometry

It was Monge who occupied a central position in the advancement of differential geometry during the late eighteenth century, following Alexis Claude Clairaut (1713–1765) and Leonhard Euler (1707–

---

[26]Stachel, 2003, p.329.
[27]Cardone, 2003, p. 74.

1783).²⁸ He pioneered the subject and his process of thinking about differential geometry was distinctive. Monge used a combination of geometric reasoning, reinforced by the tools of calculus and analytic geometry of space, that emphasized the theory of differential equations in solving problems. As a modern teacher, Monge was well known for relating his ground-breaking mathematics to students in this way. This theme is prevalent throughout his textbook, *Application de l'Analyse à la Géométrie* (1807) which he taught from at the École Polytechnique, see Figure 3.6. It was the first treatise on differential geometry of curves and surfaces. In fact, geometric studies in France made the greatest progress during Monge's prime and in the first decades of the nineteenth century.²⁹

---
²⁸Kolmogorav, 1996, p. 5.
²⁹*Ibid.*

> APPLICATION
> # DE L'ANALYSE
> ## A LA GÉOMÉTRIE,
> A L'USAGE
> DE L'ÉCOLE IMPÉRIALE POLYTECHNIQUE;
> PAR M. MONGE.
>
>
>
> PARIS,
> BERNARD, Libraire de l'École Impériale Polytechnique
> et de l'Ecole Impériale des Ponts et Chaussées, Editeur des
> Annales de Chimie, quai des Augustins, n°. 25.
> 1807.

Figure 3.6: Title Page of *Application de l'Analyse à la Géométrie* (1807)

Application de l'Analyse à la Géométrie is brimming with numerous geometrical queries, in-depth analytical discussions, and compelling applications on surfaces and space curves. Monge develops each concept by posing questions and providing elementary formulas as a basis for the solution, not according to the usual presentation of hypothesis, claim, and proof. His work is "marked by its linking the most vivid spatial intuition with analytical operations in the most natural way. An analytic formula appears not as an end itself, but only as the shortest expression to actually perceived spatial relations; its further development is based on spatial constructions."[30]

3.2.1 Examples from *Application de l'Analyse à la Géométrie*

The lessons Monge included in his textbook were carefully crafted to illustrate a wide variety of concrete problems that may be considered when studying curves and surfaces in space. In nearly all of his discussions in the book, Monge starts out geometrically; he describes the surface with lucid precision, then translates the visual description into various equations and analytic expressions, often leaving the reader to fill in the mathematical details between each step. The following examples from Monge's textbook highlight the explicitly descriptive geometrical reasoning that he used to elucidate some of the principal concepts from differential geometry in three-dimensions. The selections presented here were chosen for their applicability to the teaching of Monge's ideas in a present-day differential geometry course. Specific curves and surfaces given in each problem, as Monge envisioned them, are recreated and conceptualized using the mathematical computer algebra and graphics software program, Maple.

3.2.2 Involutes, Evolutes and Radius of Curvature of Space Curves

In 1785, Monge wrote "Mémoire sur les développées, les rayons de courbure, et les différens genres d'inflexions des courbes à

---

[30]Klein, as translated by R. Hermann, 1979, p. 71.

double courbure."[31] This publication is included as Chapter XXVII of *Application de l'Analyse à la Géométrie* (1807). In the paper, Monge introduces the concept of the curvature of a space curve, what he calls *le rayon de courbure* and he distinguishes between two types of *points d'inflexion*: inflexion developed by zero torsion and inflexion developed by zero curvature. Monge had the ideas of curvature and torsion in mind, but he did not specifically mention the term *torsion* or give an explicit analytic expression for the concept of torsion. The analysis that follows uses Monge's ideas on the development of involutes, evolutes, and radius of curvature. These concepts are written in terms of curvature and torsion, as they are defined using conventional mathematical notation. Their geometrical forms in two and three dimensions are rendered using Maple.

In defining *le rayon de courbure* or the radius of curvature of a space curve, Monge begins with a geometric description of how *développantes,* i.e., involutes, of plane curves are formed.[32] This is a striking example of how Monge is able to transition and perpetuate his earlier work from *Géométrie Descriptive* (1799) to new problems in differential geometry. Figure 3.7 shows Monge's construction of an involute referenced in Planche XXIII, Fig. 43 of *Géométrie Descriptive* (1799) in which the tangents to the plane curve $MNP'O$ sweep out the involute curve $GPP'P''H$.

---

[31]See Appendix A for a list of *The Mathematical and Scientific Works of Gaspard Monge.*

[32]Monge, *Géométrie Descriptive,* 1799, p. 106, Section V, No. 104.

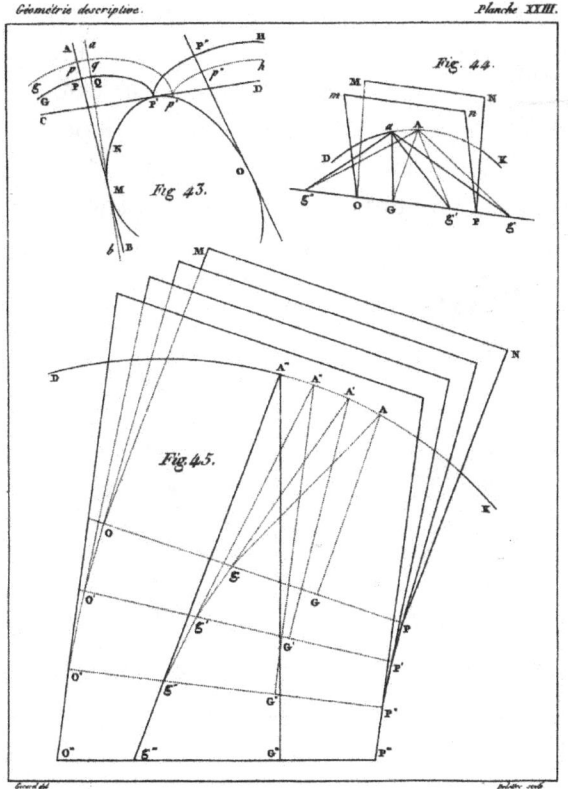

Figure 3.7: Monge's construction of a plane curve and its involute from *Géométrie Descriptive* (1799)

Monge declares that $MNP'O$, called the *développée* or evolute, has an infinite number of different involutes, e.g., another is $gpp'p''h$, all of which have the same normal line. However, for plane curves it is not true that each involute has an infinite number of evolutes.

To draw a mathematically accurate involute, it is convenient to use a two or three-dimensional curve $\alpha(t)$ in parametric form. Let $\alpha(t)$ be a plane curve,

$$\alpha : (a, b) \to \mathbb{R}^2,$$

attach a string to a fixed point on $\alpha(t)$ and wind the string around the curve. If the string is pulled taut as it is unwound from the curve, the part of the string that has been unwound will be equal to the length of the curve $s(t)$ from the fixed point. In addition, the taut string will be tangent to the curve at each point in the direction of $-\alpha(t)$. The curve traced out as the string is unwound is called the *involute*. The involute

$$\beta : (a, b) \to \mathbb{R}^2$$

is given by the equation

$$\beta(t) = \alpha(t) - s(t) \frac{\alpha'(t)}{|\alpha'(t)|}$$

where

$$s(t) = \int_c^t |\alpha'(u)| du$$

is the arc length with c the starting point of $\alpha(t)$.[33] The stipulation that the string is taut indicates that $\beta'(t)$ is perpendicular to $\alpha'(t)$ for all values of t, that is, the curve $\beta(t)$ intersects the generating tangent lines of $\alpha(t)$ at right angles. As $\beta(t)$ is called an *involute* of $\alpha(t)$, curve $\alpha(t)$ is called the *evolute* of $\beta(t)$.[34]

---

[33]Note that if $\alpha(t)$ is a unit-speed curve parameterized by arc length, s, then, $|\alpha'(s)| = 1$ and

$$s(t) = \int_c^t |\alpha'(u)| du = t - c.$$

[34]McCleary, 1993, p. 73. In his efforts to build an accurate pendulum clock, Christiaan Huygens (1629–1695) was the first to study involutes and evolutes. His work, *Horologium Oscillatorium sive de motu pendulorum* (1673), includes the result that if a pendulum is built with a flexible string constrained to unwind along a fixed curve α, the pendulum follows a path determined by an *involute* of α. It is because of Huygens that we know in order to build an ideal pendulum clock, we must construct the *evolute* of a cycloid. An application of

For example, the parameterized unit circle
$$\alpha(t) = (\cos t, \sin t)$$
with starting point
$$\alpha(t)|_{t=0} = \alpha(0) = (1, 0)$$
has an involute given by
$$\beta(t) = (\cos t, \sin t) - s(t) \frac{(-\sin t, \cos t)}{\sqrt{\sin^2 t + \cos^2 t}}$$
$$= (\cos t, \sin t) - s(t)(-\sin t, \cos t)$$
where
$$s(t) = \int_0^t \sqrt{\sin^2 u + \cos^2 u}\, du = \int_0^t du = t.$$
Thus,
$$\beta(t) = (\cos t, \sin t) - t(-\sin t, \cos t)$$
$$= (\cos t + t \sin t, \sin t - t \cos t).$$

---

circle involutes is an involute gear whose tooth profile is based on an involute curve. It was designed in the 18th century by Euler. This type of gear is popular even today because of how well its teeth mesh, wasting very little space. The two papers Euler wrote on gear teeth are: "On finding the best shape for gear teeth.", Eneström number E249, and "Supplement on the shape of the teeth of wheels [gear teeth]", E330.

```
circle := plot([cos(t), sin(t),
    t = 0 .. 2*Pi],
    color = black, thickness = 2):

involute := plot([cos(t)+t*sin(t),
                sin(t)-t*cos(t),
    t = 0 .. 2*Pi],
    color = brown, thickness = 3):

display({circle, involute},
    scaling = constrained);
```

Listing 3.1: Circle with one involute

Two Maple plots of a circle and its involutes are given in Figures 3.8 and 3.9. Visually, it can be deduced from Figure 3.9 that an infinite number of other involutes of the same circle may be drawn if we begin unwinding the thread at different points on the circumference. Two variations of Maple code for computing involutes as shown in Figures 3.8 and 3.9 are included. Listing 3.1 shows a direct computation of the involute, specifically for the curve $\alpha(t) = (\cos t, \sin t)$. Listing 3.3 provides a general Maple procedure for calculating the formula and plotting the graph of an involute for any curve $\alpha(t)$ parameterized by t.[35]

---

[35]Maple was used throughout this book. When opening a new Maple file, the following commands are necessary to execute the given code, **restart**: **with(plots)**: **with(VectorCalculus)**:. In setting up the Maple procedure beginning with the command, **proc**: the first line, **involute:=proc(curve)** gives the procedure a title and indicates the function $\alpha(t)$ to be called. The second line, **local** defines the variables to be used inside the procedure. The subsequent lines are statements for the arguments needed to implement the procedure. The last line, **end**: indicates the completion of the **proc**. To execute the involute **proc**, define a curve $\alpha(t)$ via a name and formula, then

```
circleplot := plot([
    cos(t), sin(t),
    t = 0 .. 2*Pi],
    color = black, thickness = 2):

circleinvoluteplot1 := plot([
    cos(t)+t*sin(t),
    sin(t)-t*cos(t),
    t = 0 .. 2*Pi],
    color = brown, thickness = 3):

circleinvoluteplot2 := plot([
    cos(t)+(t-1)*sin(t),
    sin(t)-(t-1)*cos(t),
    t = (1/2)*Pi - 0.9 .. 2*Pi],
    color = blue, thickness = 4):

circleinvoluteplot3 := plot([
    cos(t)+(t-2)*sin(t),
    sin(t)-(t-2)*cos(t),
    t = (1/2)*Pi - 0.9 .. 2*Pi],
    color = blue, thickness = 4):

circleinvoluteplot4 := plot([
    cos(t)+(t-4)*sin(t),
    sin(t)-(t-4)*cos(t),
    t = (1/2)*Pi - 0.9 .. 2*Pi],
    color = red, thickness = 5):

display({circleplot, circleinvoluteplot1,
    circleinvoluteplot3, circleinvoluteplot4},
    scaling = constrained);
```

Listing 3.2: Circle with several involutes

```
involute := proc(curve)
    local curveprime, curvenorm, s;
    curveprime := map(diff, curve, t);
    curvenorm := Norm(curveprime);
    subs(t = u, curveprime);
    s := simplify(int(curvenorm, u = 0 .. t));
    simplify(curve-s*curveprime/curvenorm);
end:

circle := <(cos(t), sin(t), 0>
involute(circle)

circleplot := SpaceCurve(circle,
    t = 0 .. 2*Pi,
    color = black, thickness = 2):

involuteplot := SpaceCurve(involute(circle),
    t = 0 .. 2*Pi,
    color = red, thickness = 2):

display({circleplot, involuteplot},
    scaling = constrained, axes = normal,
    orientation = [-90, 0]);
```

Listing 3.3: Involute of a curve

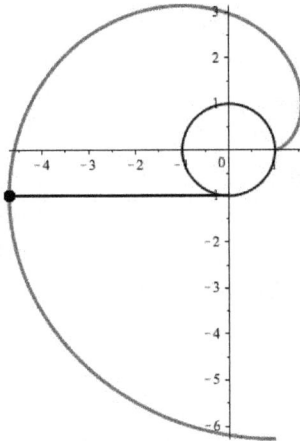

Figure 3.8: Circle and an involute

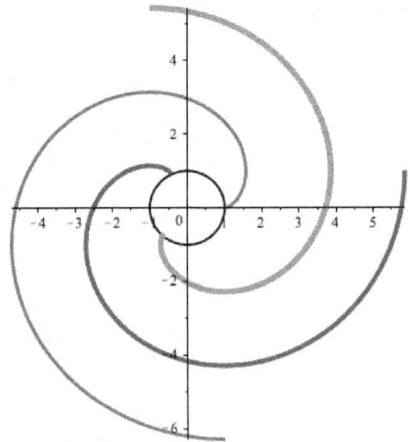

Figure 3.9: Circle and three involutes

use the command, involute(name) to compute the involute. It is assumed that $\alpha(t) = (x(t), y(t), z(t))$ is a curve parametrized by t in three dimensions, with $z(t) = 0$ for plane curves.

Monge also makes note that one can easily conceive of how the evolute is formed from a given involute, since all the normal lines of the involute are tangent to the evolute. Therefore, as shown in Figure 3.10, by taking the normals to the curve OPQ through the points P and Q, the original circle is the evolute as it is tangent to all the normals. If we let M be the point on the evolute where normals of P and Q intersect, then PM will be regarded the radius of the circle with center M, called the *osculating circle*.[36] Thus, the circle centered at M has the same curvature as the arc PQ. Monge calls the radius PM, *le rayon de courbure* or the radius of curvature of the plane curve. The point M where the two consecutive normals meet is the *le centre de courbure* or center of curvature. From this it can be concluded that the centers of the osculating circles to $\alpha(t)$ lie on the evolute curve.

---

[36]Struik, 1950, p. 419. Leibinz first introduced the term *circle of osculation*. As the tangent line best approximates a curve at a point P, the osculating circle (also, circle of curvature) is the circle that best approximates the curve at P. The osculating circle is tangent to a curve at a point such that it is a limit of the circle formed from three neighboring points on the curve, i.e., it has contact of at least order two with the curve.

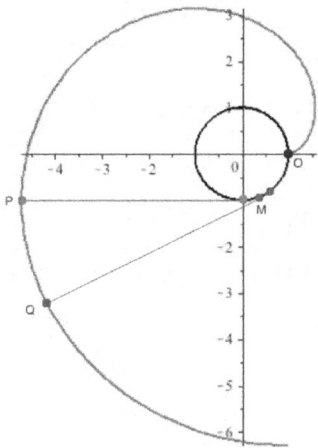

Figure 3.10: Constructing the evolute of a plane curve

The evolute of a plane curve with involute $\beta(t)$ is written as

$$\varepsilon(t) = \beta(t) + \frac{1}{\kappa(t)} N(t)$$

where

$$\kappa(t) = \frac{T'(t)}{|\beta'(t)|} = \frac{|\beta'(t) \times \beta''(t)|}{|\beta'(t)|^3}$$

is the curvature and

$$N(t) = \frac{T'(t)}{|T'(t)|}$$

is unit normal of $\beta(t)$, assuming $\beta(t)$ is regular curve, i.e., $\beta'(t) \neq 0$ for all t in an interval.

As an example, suppose it is required to find the evolute of the parabola

$$y = \frac{x^2}{4}$$

with

$$\beta(t) = \left(t, \frac{t^2}{4}\right).$$

First, we determine
$$\beta'(t) = \left(1, \frac{t}{2}\right), \quad \text{with} \quad |\beta'(t)| = \frac{\sqrt{4+t^2}}{2},$$
and
$$T(t) = \frac{\beta'(t)}{|\beta'(t)|} = \left(\frac{2}{\sqrt{4+t^2}}, \frac{t}{\sqrt{4+t^2}}\right).$$
From this, we compute
$$\kappa(t) = \frac{4}{(4+t^2)^{3/2}}$$
and
$$N(t) = \left(\frac{-t}{\sqrt{4+t^2}}, \frac{2}{\sqrt{4+t^2}}\right)$$
so that the evolute of $\beta(t)$ is
$$\varepsilon(t) = \left(t, \frac{t^2}{4}\right) + \frac{(4+t^2)^{3/2}}{4}\left(\frac{-t}{\sqrt{4+t^2}}, \frac{2}{\sqrt{4+t^2}}\right)$$
$$= \left(-\frac{t^3}{4}, \frac{3t^2}{4} + 2\right).$$

A Maple rendering of the parabola and its evolute is displayed in Listing 3.4 with output shown in Figure 3.11.

```
evolute := proc (curve)
    local k, N;
    k := simplify(Curvature(curve));
    N := PrincipalNormal(curve, normalized);
    simplify(curve+N/k);
end:

parabola := <t, (1/4)*t^2, 0>
evolute(parabola)

PVevoluteparabola := PositionVector([
    evolute(parabola)[1],
    evolute(parabola)[2],
    evolute(parabola)[3]]):

parabolaplot := SpaceCurve(parabola,
    t = -6 .. 6,
    color = blue, thickness = 2):

evoluteparabola := SpaceCurve(PVevoluteparabola,
    t = -3 .. 3,
    color = black, thickness = 3):

display({evoluteparabola, parabolaplot},
    scaling = constrained,
    orientation = [-90, 0, 0]);
```

Listing 3.4: Evolute of a curve

It should be noted in the Maple procedure for generating the evolute, the curve $\beta(t)$ is parameterized by t and given using three coordinates. For plane curves, the third coordinate is assigned to be zero.

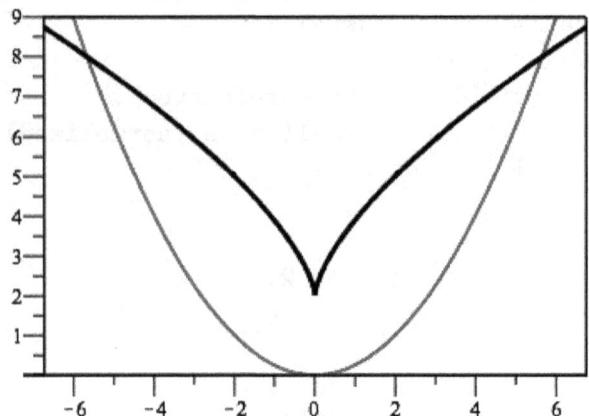

Figure 3.11: Parabola and its evolute

Figure 3.12 shows several figures of the constructions Monge drew in his *Application de l'Analyse à la Géométrie* to determine an evolute of a space curve. His drawings and theoretical explanations reveal that evolutes of space curves differ from those of plane curves; a space curve may have many evolutes, whereas, a plane curve has a unique evolute.[37] For example, consider the diagram in the lower right corner of Figure 3.12 (labeled Fig. 6 of Pl. III). In this figure, $BB'B''B'''B''''$ is a given space curve. Monge explains how each point of this curve produces two tangents BO and BP, $B'O'$ and $B'P'$, etc. which are rulings on the *tangent developable surface*.[38] That is, a tangent developable surface is a ruled surface generated by the tangent lines to a space curve.

---

[37]Monge, *Application de l'Analyse à la Géométrie*, 1807, p. 362, Chapter XXVII, Section XXVIII.

[38]Charles Tinseau (1749-1822), a premiere pupil of Monge at the military school at Mézières, was among the first to consider the osculating plane and the developable surface generated by tangents (previously made known by Clairaut).

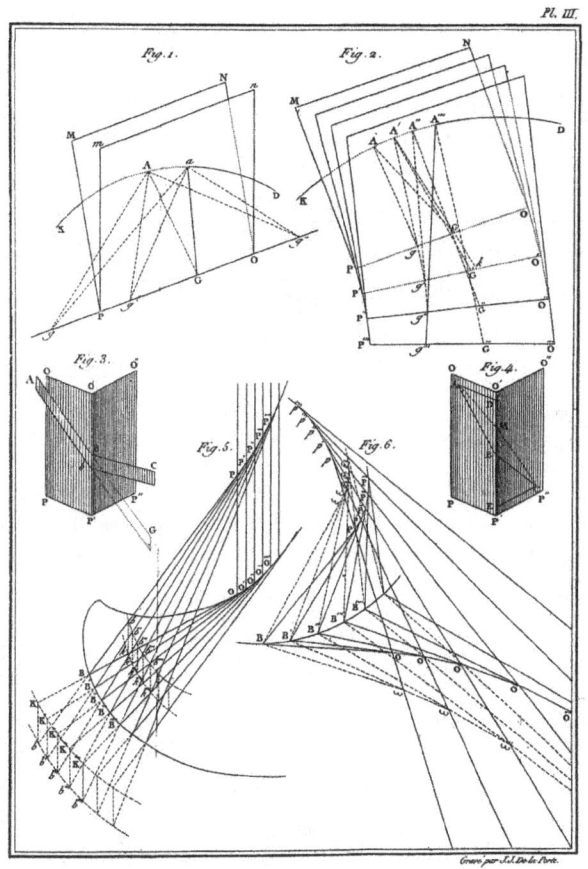

Figure 3.12: Monge's construction of an evolute of a space curve from *Application de l'Analyse à la Géométrie* (1807)

Using each pair of tangents and formulas presented earlier in the lesson, Monge produces two distinct evolutes $OO'O''O'''$ and $PP'P''P'''$ to the original space curve $BB'B''B'''B''''$.

In Monge's exposition of the theory of space curves, he mathematically constructs the ideas of *radius of curvature, developable surfaces,* and *osculating circle.* Figure 3.13 details the construction

75

of a space curve and an evolute using a method similar to Monge's own. An analysis of this figure depicts the space curve MM_1M_2 with normal planes MCN, $M_1C_1N_1$, and $M_2C_2N_2$. The lines CP, C_1P_1, and C_2P_2 are the lines of intersection of two consecutive normal planes which Monge calls *le lieu des poles* or the *polar lines*. These lines are the generators of the *polar developable* surface, and therefore are tangents to the cuspidal edge PP_1P_2 called *l'arête de rebroussement* or the *edge of regression* of that surface. Thus, the *polar developable* is the surface generated by, or as Monge states, enveloped by, the normal planes to a space curve. When the line EM is drawn arbitrarily and the lines E_1M_1 and E_2M_2 are constructed in such a way that the points E, E_1, and E_2 meet the generators (polar lines) CP, C_1P_1, and C_2P_2 respectively, the lines EM, E_1M_1, and E_2M_2 (which are tangents to the curve EE_1E_2) are normal to the curve MM_1M_2. This makes the curve EE_1E_2 traced on the polar developable an evolute of the given space curve MM_1M_2. To visualize this further, if EM is part of the thread wound around curve EE_1E_2, then as the thread is unwound, the point M will move along the given curve MM_1M_2. Monge points out that since the first line EM was arbitrary, the curve has an infinite number of evolutes all lying on the polar developable surface[39]

Furthermore, given the space curve MM_1M_2, the small straight line segments MM_1 and M_1M_2 represent the tangent lines to the space curve. The plane containing two consecutive tangent lines is called the *osculating plane*. In Figure 3.13, the osculating plane

---

[39]In fact, Monge showed that all of the evolutes of a space curve are geodesics. A curve is a geodesic when two successive tangents to it make equal angles with the intersection of the corresponding tangent planes of the surface. Also, a geodesic is the shortest distance between two points on a curved surface, which turns out to be the path of least curvature). In relation to Figure 3.13, EM and E_1M_1 are two successive tangents to the evolute EE_1E_2 that form equal angles with the line CP, the intersection of two consecutive tangent planes of the polar developable.

MM_1M_2 meets the first two normal planes in the lines MC and M_1C. Monge calls the point C the first center of curvature, since it is the center of the *osculating circle* of which MM_1 is a small arc. Similarly, the second center of curvature, C_1, is constructed by the intersection of the lines $M_1 C_1$ and $M_2 C_1$ in which a second osculating plane $M_1M_2M_3$ intersects the second and third normal planes. Monge calls M_1C and M_1C_1 the *radii of curvature*.

To summarize the constructions in Figure 3.13, the space curve is the locus of the points M, M_1, M_2; the points C, C_1, C_2,... are the corresponding *centers of curvature* (i.e., the centers of the *osculating circles*); the planes MCN, $M_1C_1N_1$, ... are normal to the space curve; the lines CP, C_1P_1, ... are the *polar lines*; and the points P, P_1, P_2 are the centers of the *osculating spheres*. The osculating sphere passes through four consecutive points of the curve. Another way to describe this sphere is to consider any space curve and a point P on the curve such that the point P is on the normal line to the osculating plane at the center of curvature. Next, consider a moving sphere whose center is at the point P, this sphere intersects the osculating plane in the osculating circle. Hence, the sphere is called the osculating sphere and the *edge of regression* of the polar developable surface is the locus of the centers of the osculating spheres. In general, two other curves may be determined from a given space curve, the locus of the centers of curvature of the osculating circle and the locus of the centers of spherical curvature of the osculating spheres. It is also of importance to note that an evolute of a space curve may be the locus of the centers of the osculating circles, or osculating spheres, or neither.

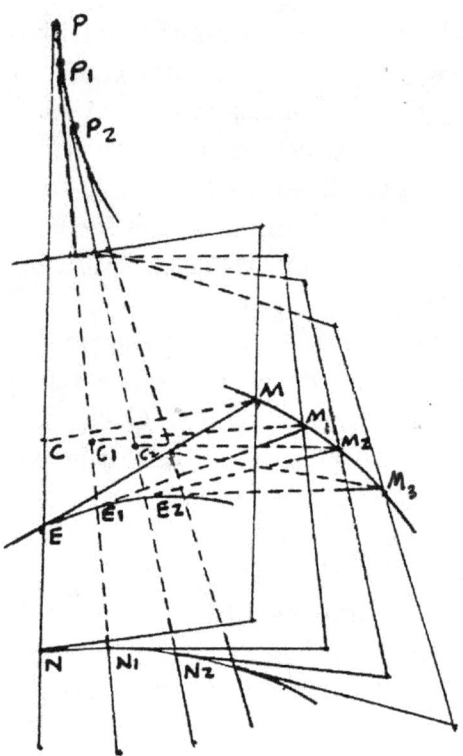

Figure 3.13: Construction of an evolute to a space curve

In Chapter XXVII, Section XXII of *Application de l'Analyse à la Géométrie*, Monge sets out to determine the *edge of regression* of the polar developable surface. In translation, his discussion proceeds as follows:

> Let
> $$x, \quad y = \varphi(x) \quad \text{and} \quad z = \psi(x)$$
> be the components of a space curve. The normal plane at the point
> $$(u, \varphi(u), \psi(u))$$

is given by

$$(z - \psi(u))\psi'(u) + (y - \varphi(u))\varphi'(u) + (x - u) = 0 \quad (1)$$

Differentiating and simplifying, we have

$$(z - \psi(u))\psi''(u) + (y - \varphi(u))\varphi''(u) - \\ \left[1 + (\varphi'(u))^2 + (\psi'(u))^2\right] = 0 \quad (2)$$

Differentiating once again and combining like terms gives us

$$(z - \psi(u))(\psi'''(u) + (y - \varphi(u))\varphi'''(u) - \\ 3\left[\varphi'(u)\varphi''(u) + \psi'(u)\psi''(u)\right] = 0 \quad (3)$$

It is only necessary to eliminate u from equations (1), (2), and (3), and the two equations in x, y, and z that are obtained, will be the *edge of regression* that we are seeking.[40]

To better understand and conceptualize the outcome of Monge's work, current notation and theory from differential geometry will be utilized to find the equation of the evolute of a space curve. The equations will be twice differentiated, as Monge did, to obtain the desired result.

Given a space curve $c(s)$ parameterized by arc length s. By definition, the tangents of the evolute to $c(s)$ are normal to $c(s)$, thus the equation of the evolute of a space curve must be of the form

$$E = c(s) + aN + bB \quad (4)$$

To find a and b, we use the fact that

$$(E - c(s)) \cdot T = 0 \quad (5)$$

---

[40] Monge, *Application de l'Analyse à la Géométrie*, 1807, p. 357, Chapter XXVII, Section XXII.

where E is independent of s, and c and T are functions of s. The formula in (5) gives the equation of the normal plane of a space curve in vector form. Monge's equation for the normal plane is given in (1). Differentiating (5), we obtain

$$(E - c(s))' \cdot T + (E - c(s)) \cdot T' = 0$$

where we can conclude the following using the Frenet-Serret formulas[41]

$$(E - c(s))' = -T \quad \text{and} \quad T' = \kappa N.$$

The parameter $\kappa = \kappa(s)$ is the curvature of $c(s)$ parameterized by arc length s. We can now write

$$\begin{aligned} -T \cdot T + (E - c(s)) \cdot \kappa N &= 0 \\ -1 + (E - c(s)) \cdot \kappa N &= 0 \\ (E - c(s)) \cdot \kappa N &= 1. \end{aligned} \qquad (6)$$

Next, from (4) we have

$$(E - c(s)) = aN + bB$$

so that the last equation in (6) gives us

$$\begin{aligned} (aN + bB) \cdot \kappa N &= 1 \\ (aN \cdot \kappa N) + (bB \cdot \kappa N) &= 1 \\ a\kappa + 0 &= 1 \\ a &= \frac{1}{\kappa} \end{aligned}$$

---

[41]Struik, 1969, p. 19. These formulas are expressions relating the curvature and torsion to the moving orthogonal frame of unit vectors T, N, and B along a space curve. J.F. Frenet (1816–1900) originally obtained the results in his dissertation of 1847, but they were also published in a paper by J.A. Serret (1819–1885) in 1851 before Frenet's results were well known.

To simplify the above equations the formulas, $N \cdot N = 1$ and $B \cdot N = 0$ were used.

The value of a makes sense since we know that the radius of the osculating circle in the direction of the normal to the curve at a point on the curve is $\frac{1}{\kappa}$. To find b, we simplify the last equation in (6) a bit more, then differentiate:

$$(E - c(s)) \cdot \kappa N = 1$$
$$\kappa(E - c(s)) \cdot N = 1$$
$$(E - c(s)) \cdot N = \frac{1}{\kappa}$$

differentiating, we obtain

$$(E - c(s))' \cdot N + (E - c(s)) \cdot N' = (\tfrac{1}{\kappa})' \qquad (7)$$

Substitute

$$(E - c(s))' = -T \quad \text{and} \quad N' = -\kappa T + \tau B$$

into (7) and simplify.[42]

$$-T \cdot N + (E - c(s)) \cdot (-\kappa T + \tau B) =$$
$$(E - c(s)) \cdot (-\kappa T) + (E - c(s)) \cdot (\tau B) =$$
$$(aN + bB) \cdot (-\kappa T) + (E - c(s)) \cdot (\tau B) =$$
$$aN \cdot (-\kappa T) + bB \cdot (-\kappa T) + (E - c(s)) \cdot (\tau B) =$$
$$0 + 0 + (E - c(s)) \cdot (\tau B) =$$
$$(E - c(s)) \cdot (\tau B) = \left(\frac{1}{\kappa}\right)'$$

Thus, we have

$$(E - c(s)) \cdot B = \frac{1}{\tau \kappa'} \qquad (8)$$

---

[42] The formula $T \cdot N = 0$ was used in the simplification. The parameter τ is the *torsion* of the curve.

and then,[43]
$$(aN + bB) \cdot B = \frac{1}{\tau \kappa'}.$$

Recalling that $N \cdot B = 0$, $B \cdot B = 1$, and $a = \frac{1}{\kappa}$ we have

$$b = \frac{1}{\tau \kappa'}.$$

The edge of regression of the polar developable surface of $c(s)$ is finally

$$E = c(s) + \frac{N}{\kappa'} + \frac{B}{\tau \kappa'}.$$

As the edge of regression is the curve formed by the locus of the centers of curvature of the osculating circles or spheres, in some cases it is also an evolute of the space curve.

When Monge introduced the concept of curvature in his 1785 paper, "Mémoire sur les développées, les rayons de courbure, et les différens genres d'inflexions des courbes à double courbure", he distinguished between two types of *pointes d'inflexion* of a curve. In general, Monge identifies a *point of inflection* as the point where the curve changes concavity. Monge states that the curve "loses

---

[43]Equation (8) holds for unit speed curves $c(s)$ for which the curvature is parameterized by arc length, $\kappa = \kappa(s)$. Since

$$\left(\tfrac{1}{\kappa}\right)' = -\tfrac{\kappa'}{\kappa^2}$$

and

$$\kappa'(t) = \kappa'(s(t)) \cdot s'(t)$$

we have

$$\kappa'(s) = \tfrac{\kappa'(t)}{s'(t)} = \tfrac{\kappa'(t)}{\nu(t)}.$$

Thus, for a non-unit speed curve $c(t)$, we can define

$$\tfrac{1}{\tau \kappa'} = -\tfrac{1}{\tau} \cdot \tfrac{\kappa'(t)}{|\nu(t)|\kappa(t)^2}.$$

its curvature" at this point and that two consecutive straight-line elements (or what he calls infinitesimally short chords, one on either side of the point) form a straight line. He continues to inform that a space curve may have two kinds of inflexion, the first are called *points de simple inflexion* which occur when four nearby points of the curve lie in one plane, i.e., three consecutive straight line elements lie in the same plane. The second are called *points de double inflexion* and they occur when three consecutive points, which form two consecutive straight line segments, lie on a straight line. The *points de simple inflexion* are those points determined by zero torsion and the *points de double inflexion* are points when the curvature is zero. Monge asserts, that at an inflection point of a space curve, either the curvature or the torsion vanishes. It was not until the 19th century that curvature and torsion were regarded as the two fundamental properties of a space curve. It was proven that if the curvature and torsion are expressed in terms of arc length, then the curve and its position in space may be completely determined.

3.2.3 Maple Exposition of Space Curves, Evolutes, Osculating Circles, Osculating Spheres, and the Polar Developable Surface

Differential geometry in three dimensions is one of the most intuitive areas of advanced mathematics. Mathematical software programs capable of performing complex computations and visualizations, make it possible to combine the traditional formal treatment of the subject matter with a strong hands-on and intuitive approach in teaching the subject. These programs enable us to actually see the mathematics we are talking about. We now turn to the computer algebra system, Maple, to bring to life several applications from the theory of curves and surfaces found in Monge's *Application de l'Analyse à la Géométrie.*

Helix

A simple parametrization of the helix is given by

$$c(t) = (a\cos t, a\sin t, bt)$$

in which a is the radius of the helix and b is the slope of the helix. This curve is of interest as its curvature and torsion are both constant.

Using the Maple involute procedure found in Listings 3.3 and 3.5, we discover that the involute of the helix with $a = b = 1$ is the recognizable planar curve given in Figure 3.14. Listing 3.6 creates a plot of a helix and an evolute using the brute force method in Maple. In Figure 3.15, the helix with $a = .8$, $b = .6$ and an evolute are shown. The concise procedure for computing and plotting the evolute of a curve is then given in Listing 3.7.

```
helix := <cos(t), sin(t), t>
involute(helix)

helixplot := SpaceCurve(helix,
    t = 0 .. 4*Pi,
    color = black, thickness = 2):

involutehelix := SpaceCurve(involute(helix),
    t = 0 .. 4*Pi,
    color = brown, thickness = 3):

display({helixplot, involutehelix},
    scaling = constrained,
    orientation = [-69, 77, 0]);
```

Listing 3.5: Involute of a helix

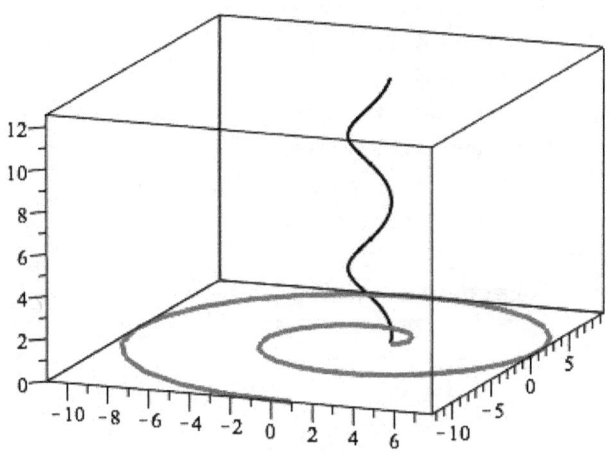

Figure 3.14: Helix with its involute

```
x:=t->0.8*cos(t):
y:=t->0.8*sin(t):
z:=t->0.6*t:
c:=[x(t),y(t),z(t)]:
n:=t->Norm(<0.8*cos(t),0.8*sin(t),0.6*t>); n(t);
T:=TangentVector(t-> <x(t),y(t),z(t)>, normalized)
    assuming t::real: T(t);
k:=Curvature(t-> <0.8*cos(t),0.8*sin(t),0.6*t>)
    assuming t::real;k(t);
simplify(k(t)) assuming t::real;
kprime:=diff(k(t),t);(k(t))^2;
N:=PrincipalNormal(t-> <x(t),y(t),z(t)>, normalized)
    assuming t::real: N(t);
B:=t->Binormal(PositionVector([x(t),y(t),z(t)],
    t,normalized) assuming t::real: B(t);
tau:=Torsion(t-><0.8*cos(t),0.8*sin(t),0.6*t)
    assuming t::real: simplify(tau(t));
evolute3d:=
  [x(t)+(N(t)[1]/k(t))-(kprime*B[1])/(n*tau*k^2),
  y(t)+(N(t)[2]/k(t))-(kprime*B[2])/(n*tau*k^2),
  z(t)+(N(t)[3]/k(t))-(kprime*B[3])/(n*tau*k^2)]:
simplify(evolute3d);
P1:=spacecurve(c,t=0..4*Pi,
    color=blue,thickness=3):
P2:=spacecurve(evolute3d, t=0..4*Pi,
    color=black, thickness=3):
display({P1,P2},
    scaling=constrained,axes=boxed);
```

Listing 3.6: Evolute of a helix

```
evolute := proc(curve)
    local k, N;
    k := simplify(Curvature(curve));
    N := PrincipalNormal(curve, normalized);
    simplify(curve+N/k)
 end:

helix := <0.8*cos(t), 0.8*sin(t), 0.6*t>
evolute(helix)

PVevolutehelix := PositionVector([
    evolute(helix)[1],
    evolute(helix)[2],
    evolute(helix)[3]]):

helixplot := SpaceCurve(helix,
    t = 0 .. 5.5*Pi,
    axes = boxed, color = blue, thickness = 2):

evolutehelix := SpaceCurve(PVevolutehelix,
    t = 0 .. 5.5*Pi,
    axes = boxed, color = brown, thickness = 4):

display({evolutehelix, helixplot},
    scaling = constrained, axes = boxed);
```

Listing 3.7: Evolute of a curve

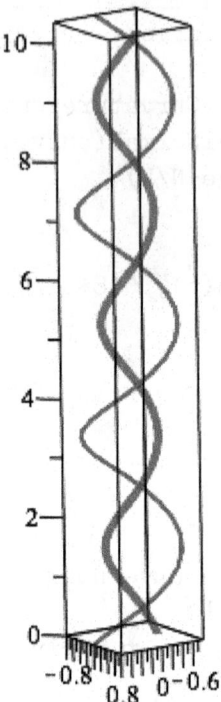

Figure 3.15: Helix with an evolute

The curvature and torsion are constant for the helix; viz. $\kappa = 0.8$ and $\tau = 0.6$; which is reflected in the output of the Maple code in Listing 3.6. Since the curvature is constant, its derivative $\kappa' = 0$. Hence, the binormal vector component B in the evolute equation

$$E = c(s) + \frac{N}{\kappa'} + \frac{B}{\tau\kappa'}$$

will vanish. The helix in this example is not a unit speed curve. Thus, the coefficient of the binormal vector is computed using the formula

$$\frac{1}{\tau\kappa'} = -\frac{1}{\tau} \cdot \frac{\kappa'(t)}{|\nu(t)|\kappa(t)^2}.$$

This means that the only contributing component is the normal vector. To find a point on the evolute we go out a distance equal to the radius of curvature, $\frac{1}{\kappa}$, from a point on the space curve in the direction of the normal. The circle that best approximates a curve at a point t_0 is called the *osculating circle*. As we would suspect, the centers of the osculating circles lie on the evolute of the helix.

The center of the osculating circle of a curve $c(t)$ at the point t_0 is

$$p = c(t_0) + \frac{1}{\kappa(t_0)} N(t_0).$$

The equation of the osculating circle with center $p(t_0)$ is

$$C = p(t_0) + \frac{1}{\kappa(t_0)} N(t_0) \cos t + \frac{1}{\kappa(t_0)} T(t_0) \sin t.$$

In Figure 3.16, a plot of the helix

$$c(t) = (0.8 \cos t, 0.8 \sin t, 0.6t)$$

its evolute, and three osculating circles for $t = \pi$, $t = \frac{5\pi}{2}$, and $t = 5\pi$ are shown. Listing 3.8 gives the Maple syntax used to generate the multiple plots.[44]

---

[44]Listings 3.8 and 3.9 require execution of the procedure in Listing 3.6 for the helix given in Figures 3.16 and 3.17.

```
helix := <cos(t), sin(t), t>

occenter := proc(curve, a)
    local k, N, ct0, kt0, Nt0;
    k := simplify(Curvature(curve));
    N := PrincipalNormal(curve, normalized);
    ct0 := simplify(subs(t = a, curve));
    kt0 := simplify(subs(t = a, k));
    Nt0 := simplify(subs(t = a, N));
    simplify(ct0+Nt0/kt0);
 end:

c1 := occenter(helix, Pi);
c2 := occenter(helix, 5*Pi/2);
c3 := occenter(helix, 5*Pi);

ocircle := proc(curve, b)
    local k, N, T, ct0, kt0, Nt0, Tt0;
    k := simplify(Curvature(curve));
    N := PrincipalNormal(curve, normalized);
    T := TangentVector(curve, normalized);
    ct0 := simplify(subs(t = b, curve));
    kt0 := simplify(subs(t = b, k));
    Nt0 := simplify(subs(t = b, N));
    Tt0 := simplify(subs(t = b, T)); <
    <occenter(curve, b)[1]
        +Nt0[1]*cos(t)/kt0+Tt0[1]*sin(t)/kt0,
    occenter(curve, b)[2]
        +Nt0[2]*cos(t)/kt0+Tt0[2]*sin(t)/kt0,
    occenter(curve, b)[3]
        +Nt0[3]*cos(t)/kt0+Tt0[3]*sin(t)/kt0>;
 end:
```

```
ocircle1 := ocircle(helix, Pi);
ocircle2 := ocircle(helix, 5*Pi/2);
ocircle3 := ocircle(helix, 5*Pi);

helixplot := SpaceCurve(helix,
    t = 0 .. 5.5*Pi,
    color = blue, thickness = 2):
PVevolutehelix := PositionVector(
  [evolute(helix)[1],
   evolute(helix)[2],
   evolute(helix)[3]]):
evolutehelix := SpaceCurve(PVevolutehelix,
    t = 0 .. 5.5*Pi,
    color = brown, thickness = 3):
occenter1 := pointplot3d(c1,
    symbol = diamond, color = purple, thickness = 4):
occenter2 := pointplot3d(c2,
    symbol = diamond, color = green, thickness = 4):
occenter3 := pointplot3d(c3,
    symbol = diamond, color = orange, thickness = 4):
oc1 := SpaceCurve(ocircle1,
    t = 0 .. 2*Pi,
    axes = boxed, color = black, thickness = 3):
oc2 := SpaceCurve(ocircle2,
    t = 0 .. 2*Pi,
    axes = boxed, color = black, thickness = 3):
oc3 := SpaceCurve(ocircle3,
    t = 0 .. 2*Pi,
    axes = boxed, color = black, thickness = 3):
display({oc1, oc2, oc3,evolutehelix, helixplot,
    occenter1, occenter2, occenter3},
    scaling = constrained);
```

Listing 3.8: Three osculating circles

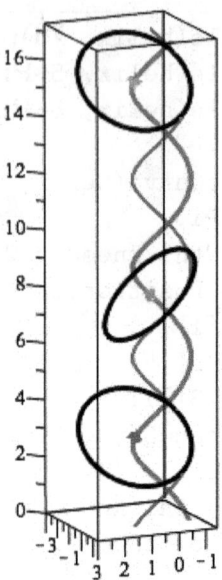

Figure 3.16: Helix with an evolute and three osculating circles

An evolute of a space curve may be the curve through all the centers of its osculating circles, or its osculating spheres, or neither. The *osculating sphere* is a sphere that passes through four consecutive points of the curve. The osculating sphere of a curve $c(t)$ at the point t_0 has center given by

$$q = c(t_0) + \frac{1}{\kappa(t_0)} N(t_0) + \frac{\kappa'(t_0)}{|c'(t_0)| - \tau(t_0)\kappa(t_0)^2} B(t_0).$$

The radius of the sphere at the point t_0 is

$$R = \sqrt{\frac{1}{\kappa(t_0)^2} + \frac{\kappa'(t_0)^2}{|c'(t_0)|^2 \tau(t_0)^2 \kappa(t_0)^4}}$$

and the osculating sphere to $c(t)$ at the point t_0 has parametrization

$$S(u, v) = (q(t_0) + R(t_0) \sin u \cos v, q(t_0) + \\ R(t_0) \sin u \sin v, q(t_0) + R(t_0) \cos u). \qquad (9)$$

In Figure 3.17, the helix, its evolute, and three osculating spheres are depicted at the points $t = \pi$, $t = 5\pi/2$, and $t = 5\pi$. As the helix is a special case of a space curve with constant curvature, we observe that the centers of its osculating spheres coincide with the centers of its osculating circles, and the trace of these points forms an evolute.

```
osphere := proc(curve, p)
    local k, kprime, n, tauw, ct0, kt0,
    kprimet0, nt0, tauwt0, R, u, v;
    k := simplify(Curvature(curve));
    kprime := map(diff, k, t); n := Norm(curve);
    tauw := simplify(Torsion(curve));
    ct0 := simplify(subs(t = p, curve));
    kt0 := simplify(subs(t = p, k));
    kprimet0 := simplify(subs(t = p, kprime));
    nt0 := simplify(subs(t = p, n));
    tauwt0 := simplify(subs(t = p, tauw));
    R := sqrt(1/kt0^2
            +kprimet0^2/(nt0^2*kt0^4*tauwt0^2));
    [occenter(curve, p)[1]+R*sin(u)*cos(v),
     occenter(curve, p)[2]+R*sin(u)*sin(v),
     occenter(curve, p)[3]+R*cos(u)];
end:

osphere1 := osphere(helix, Pi);
osphere2 := osphere(helix, 5*Pi/2);
osphere3 := osphere(helix, 5*Pi);
```

```
osphere1 := [1+2*sin(u)*cos(v),
            2*sin(u)*sin(v), Pi+2*cos(u)];
osphere2 := [2*sin(u)*cos(v),
            -1+2*sin(u)*sin(v), (5/2)*Pi+2*cos(u)];
osphere3 := [1+2*sin(u)*cos(v),
            2*sin(u)*sin(v), 5*Pi+2*cos(u)];

os1 := plot3d(osphere1,
    u = 0 .. Pi, v = 0 .. 2*Pi,
    axes = boxed, color = coral, style = wireframe):
os2 := plot3d(osphere2,
    u = 0 .. Pi, v = 0 .. 2*Pi,
    axes = boxed, color = orange, style = wireframe):
os3 := plot3d(osphere3,
    u = 0 .. Pi, v = 0 .. 2*Pi,
    axes = boxed, color = tan, style = wireframe):
display({os1, os2, os3, evolutehelix, helixplot,
    occenter1, occenter2, occenter3},
    scaling = constrained);
```

Listing 3.9: Three osculating spheres

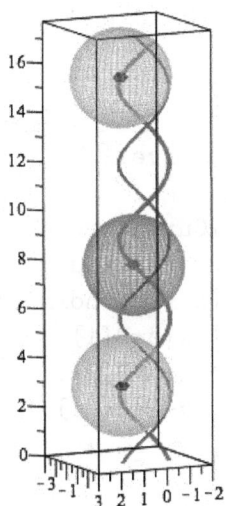

Figure 3.17: Helix with an evolute and three osculating spheres

In *Application de l'Analyse à la Géométrie*, Monge devotes considerable time discussing *polar developable surfaces* which are surfaces enveloped by the normal planes of a space curve. A general equation of a polar developable surface is given by

$$X(t,u) = c(t) + \frac{1}{\kappa}N + uB$$

We note that the distance along the normal vector N is equal to the radius of curvature $\frac{1}{\kappa}$ of the osculating circle. The parameter u in the direction of the binormal vector B is used to produce the part of the surface formed by the span of the intersecting normal planes. This is possible because the binormal vector is the normal vector to each normal plane. The polar developable surface of the helix is illustrated in Figure 3.18. It is worthy to note that the edge of regression is identified as the cuspidal edge where the two sheets of the developable surface are tangent to each other. We see that this curved edge is also the evolute of the helix. Listing

3.10 assumes that the Maple evolute procedure in Listing 3.7 was previously executed.

```
polardev := proc(curve)
    local k, N, B;
    k := simplify(Curvature(curve));
    N := PrincipalNormal(curve, normalized);
    B := Binormal(curve, normalized);
    [curve[1]+N[1]/k+u*B[1],
     curve[2]+N[2]/k+u*B[2],
     curve[3]+N[3]/k+u*B[3]];
 end:

helix := <cos(t), sin(t), t>
polardev(helix)

helixplot := SpaceCurve(helix,
    t = 0 .. 5.5*Pi,
    color = blue, thickness = 2):
PVevolutehelix := PositionVector(
    [evolute(helix)[1],
     evolute(helix)[2],
     evolute(helix)[3]]):
evolutehelix := SpaceCurve(PVevolutehelix,
    t = 0 .. 5.5*Pi,
    color = brown, thickness = 3):
pdevhelixplot := plot3d(polardev(helix),
    t = 0 .. 6*Pi, u = -2 .. 2,
    axes = boxed, shading = xy, lightmodel = light1):
display({evolutehelix, helixplot, pdevhelixplot},
    scaling = constrained);
```

Listing 3.10: Polar developable surface

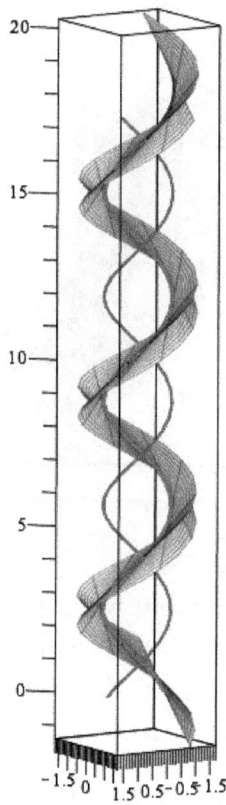

Figure 3.18: Helix with an evolute as the edge of regression of the polar developable surface

The Maple procedures written to create the new space curves and surfaces associated with the helix as shown in Figures 3.14 through 3.18 may be recreated for other curves by changing the base curve $c(t)$. This aspect is further explored in the Maple listings used to generate Figures 3.19 through 3.24.

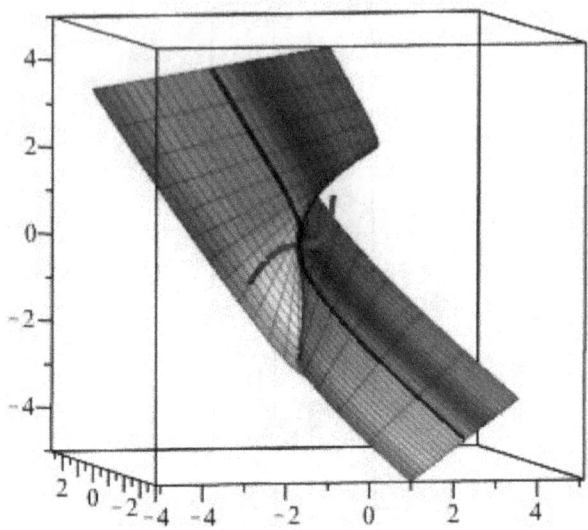

Figure 3.19: Twisted cubic with an evolute lying on the polar developable surface

Twisted Cubic

Two different evolutes of the twisted cubic

$$c(t) = (t, t^2, t^3)$$

are shown in Figure 3.19 and Figure 3.20. The evolute in Figure 3.19 lies on the polar developable surface (as it should) but it is not the edge of regression of the polar surface. In this case, the evolute curve is orthogonal to the polar lines, or rulings of the surface, everywhere on the surface except near the cusp. In Figure 3.20, a second evolute to the curve can be seen as the sharp edge where the two pieces of the surface meet, it *is* the edge of regression of the polar developable surface.

```
twistedcubic := <t, t^2, t^3>
evolute(twistedcubic)

PVevolutetwcubic := PositionVector(
    [evolute(twistedcubic)[1],
     evolute(twistedcubic)[2],
     evolute(twistedcubic)[3]]):
twcubicplot := SpaceCurve(twistedcubic,
    t = -.5 .. .5,
    color = blue, thickness = 2):
evolutetwcubic := SpaceCurve(PVevolutetwcubic,
    t = -.5 .. .5,
    color = green, thickness = 2):
display({evolutetwcubic, twcubicplot},
    scaling = constrained, axes = boxed,
    orientation = [-144, 82, 17]);

polardev(twistedcubic)

pdevtwcubicplot := plot3d(polardev(twistedcubic),
    t = -.7 .. .7, u = -3 .. .3,
    axes = boxed, color = u):
display({evolutetwcubic, pdevtwcubicplot},
    scaling = constrained);
```

Listing 3.11: Evolute and polar developable surface of twisted cubic

```
c1 := occenter(twistedcubic, 0.1);
ocircle1 := ocircle(twistedcubic, 0.1);
osphere1 := osphere(twistedcubic, 0.1);

occenter1 := pointplot3d(c1,
    symbol = diamond, color = purple, thickness = 4):
octwcubic := SpaceCurve(ocircle1, t = 0 .. 2*Pi,
    axes = boxed, color = black, thickness = 2):
osphere1 := [1+2*sin(u)*cos(v),
            2*sin(u)*sin(v), Pi+2*cos(u)]:
ostwcubic := plot3d(osphere1,
    u = 0 .. Pi, v = 0 .. 2*Pi,
    axes = boxed, style = wireframe, color = brown):
display({twcubicplot, evolutetwcubic,
        occenter1, octwcubic, ostwcubic},
    scaling = constrained);
```

Listing 3.12: Twisted cubic with osculating sphere

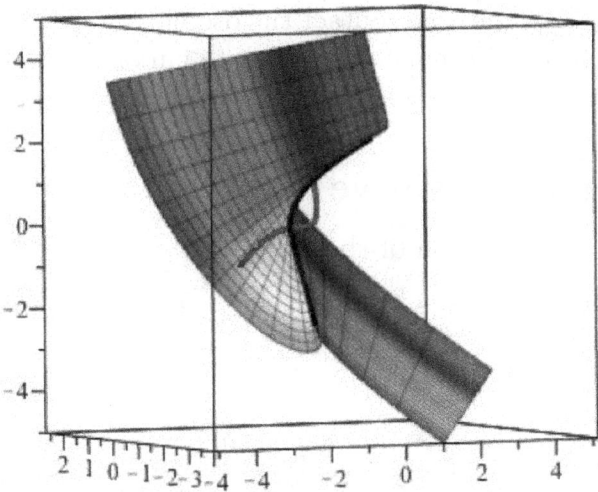

Figure 3.20: Twisted cubic with evolute as edge of regression of the polar developable surface

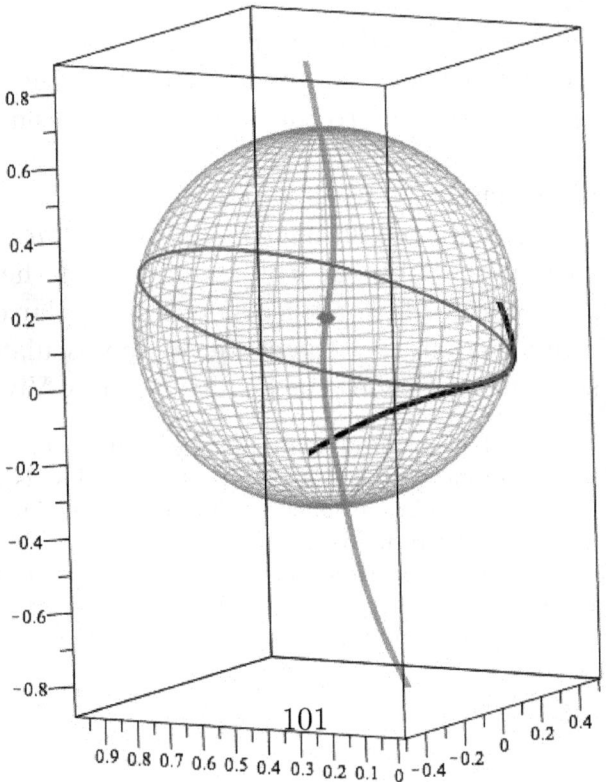

Figure 3.21: Twisted cubic on osculating sphere with

Figure 3.21 defines the contact the osculating sphere makes with the twisted cubic. Note that the evolute curve passes through the center of the sphere.

3.2.4 Viviani's Curve

The curve of intersection of the cylinder
$$(x-a)^2 + y^2 = a^2$$
and the sphere
$$x^2 + y^2 + z^2 = 4a^2$$
was studied in 1692 by Vincenzo Viviani.[45] A parametrization $c(t) = (x(t), y(t), z(t))$ of this curve, known as Viviani's curve, is
$$c(t) = \left(a(1+\cos t), a\sin t, 2a\sin\left(\tfrac{t}{2}\right)\right)$$
for $-2\pi \leq t \leq 2\pi$.

Viviani's curve is plotted in Figure 3.22. This curve presents a unique and more complex example to consider when finding an evolute of a space curve due to the fact that it lies on the surface of a sphere.

The osculating circles to Viviani's curve all have the same center as the curve has only one osculating sphere and it is *the* sphere upon which the curve lies. Subsequently, the center of the osculating sphere is fixed and the edge of regression of the polar developable surface is simply a point, i.e., the center of the osculating sphere. This makes sense when viewing the beautiful plot of Viviani's curve

---

[45]Gray, 1998. p. 201. Viviani (1622–1703) was a distinguished student of Galileo. In 1692, he proposed the following problem: "How is it possible to cut four equal windows from a hemisphere such that the remaining surface can be exactly squared?". The result is known as Viviani's curve, which is also know as Viviani's window. In addition to studying the aforementioned curve, Viviani determined the tangent to the cycloid and he conducted experiments to measure the velocity of sound relative to the firing of a cannon.

```
Viviani := SpaceCurve(<1+cos(t), sin(t),
                      2*sin((1/2)*t)>,
    t = -2*Pi .. 2*Pi,
    color = black, thickness = 4):

sphere := plot3d([2*cos(u)*cos(v),
                  2*sin(u)*cos(v), 2*sin(v)],
    u = 0 .. 2*Pi, v = -(1/2)*Pi .. (1/2)*Pi,
    style = wireframe, color = brown):

cylinder := plot3d([cos(u)+1, sin(u), v],
    u = 0 .. 2*Pi, v = -2 .. 2,
    shading = z, lightmodel = light1, color = tan):

display({Viviani, sphere, cylinder},
    scaling = constrained,
    orientation = [-38, 75, 0], axes = none);
```

Listing 3.13: Viviani's curve

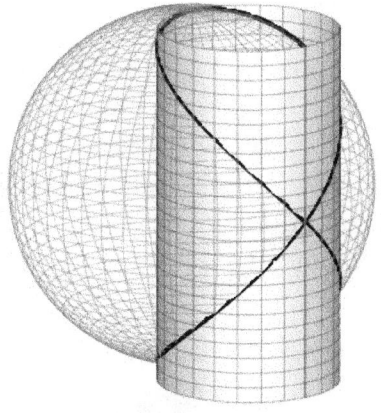

Figure 3.22: Viviani's Curve

```
evolute(Viviani)

PVevolutetViv := PositionVector([
    evolute(Viviani)[1],
    evolute(Viviani)[2],
    evolute(Viviani)[3]]):

Vivplot := SpaceCurve(Viviani,
    t = -4*Pi .. 4*Pi,
    color = black, thickness = 3):

evoluteVivplot := SpaceCurve(PVevolutetViv,
    t = -4*Pi .. 4*Pi,
    color = black, thickness = 2):

display({Vivplot, sphere, evoluteVivplot},
    scaling = constrained, axes = none,
    orientation = [-42, 87, -15]);
```

Listing 3.14: Evolute of Viviani's curve

its evolute, osculating sphere, and the polar developable surface as shown in Figure 3.24.[46] Of particular interest is the conical shape of Viviani's polar developable surface. The edge of regression of this surface is the vertex of the cone (which happens to be the center of the osculating circle). Thus, the edge of regression curve of the polar surface is not an evolute of Viviani's curve.

---

[46]See the Maple listing for the helix to create the evolute, osculating sphere, and polar developable surface for Viviani's curve.

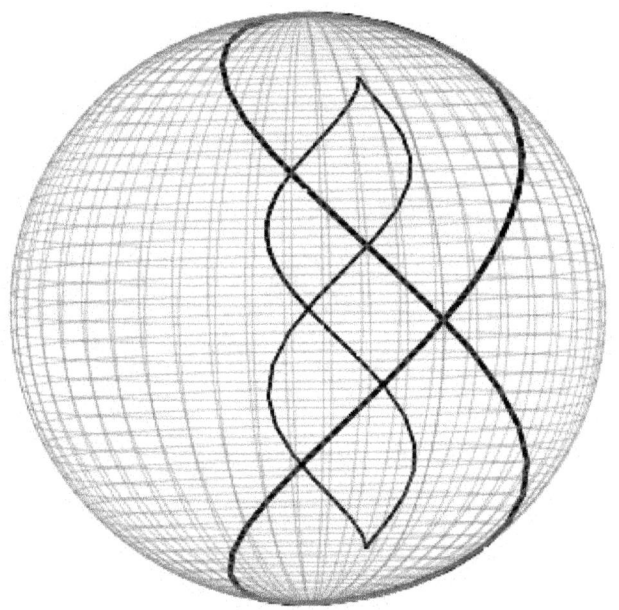

Figure 3.23: Evolute and osculating sphere to Viviani's curve

```
polardev(Viviani)

pdevViviani := plot3d(polardev(Viviani),
    t = -4*Pi .. 4*Pi, u = -3 .. 3,
    axes = none, shading = xyz,
    lightmodel = light1, numpoints = 10000):

display({Vivplot, sphere, pdevViviani},
    scaling = constrained,
    orientation = [-42, 87, -15]);
```

Listing 3.15: Polar developable surface to Viviani's curve

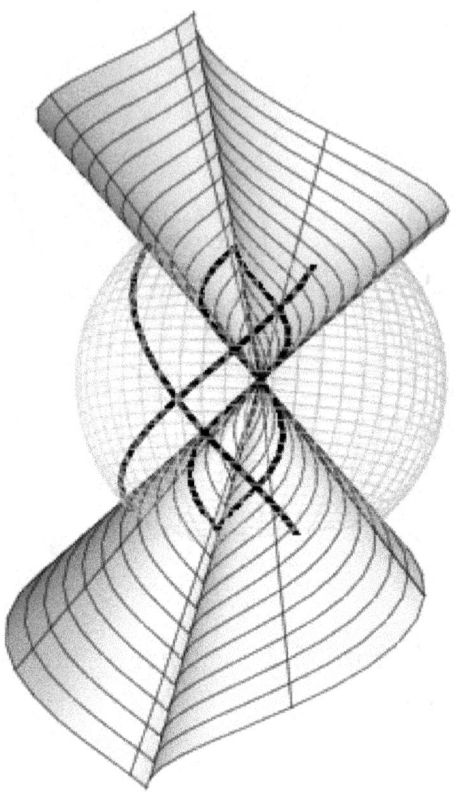

Figure 3.24: Polar developable surface and osculating sphere to Viviani's curve

3.2.5 Historical Perspective on Involutes, Evolutes, Curvature of Plane, and Space Curves

The earliest studies of plane curves resulted from those scientists and mathematicians who were attempting to build accurate and reliable clocks, which had become a necessity in the areas of scientific navigation, geography and theoretical astronomy. Christiaan Huygens (1629–1695), a Dutch astronomer, physicist, and mathematician, led the study and developed a pendulum clock that, theoretically, kept perfect time. Huygens wrote a classic volume, *Horologium Oscillatorium sive de motu pendulorum* on pendulum clocks in 1673. In his treatise, Huygens focused on the simple pendulum clock constructed by a bob connected to a string attached to a fixed point. He discovered that if the bob was forced to oscillate along a particular curve (a tautochrone) the time required to complete one oscillation would be independent of the swing amplitude.

Huygens showed that the shape of the tautochrone was a cycloid, which had been extensively investigated separately by Galileo, Mersenne, Fermat, Descartes, and Pascal and others in the seventeenth century. In his fabrication of the perfect pendulum, Huygens constructed an evolute of the cycloid and discovered that the new curve is also a cycloid.

Huygens' book is considered to be a culmination of the early theories on plane curves. His work on the evolute of a cycloid led him to develop the general theory of evolutes and involutes of plane curves, including the important concept in the early history of differential geometry, the radius of curvature.[47]

In 1731, Clairaut (1713–1765) published a book on analytic geometry of space entitled, *Recherches sur les courbes à double courbure*. At that time, a space curve was defined as an intersection of two surfaces, not as its own entity. Clairaut introduced the phrase *courbe à double courbure,* or "curve of double curvature", which became the commonly recognized terminology for a space curve.[48] Clairaut's book is regarded as the first comprehensive analysis of space curves. In 1771, Monge submitted the

---

[47]The concepts of evolute and radius of curvature date back to Apollonius of Perga (ca. 225 BCE).

[48]In fact, it was well-known hydraulic engineer and inventor of the pitot tube, Henri Pitot (1695–1771) who first used the technical name "courbes a double courbure" or "double-curvature curves" to describe space curves in a paper he wrote on the helix in 1724.

paper "Mémoire sur les développées, les rayons de courbure, et les différens genres d'inflexions des courbes à double courbure" to the Académie des Sciences. It was his first original work of great consequence. The paper was not published until 1785. He extended the geometric constructions of evolutes of plane curves from descriptive geometry, in a new direction, to curves in three-dimensions. Monge developed geometric expressions for evolutes of space curves, which led to the definition of the radius of curvature of space curve.

The study of space curves initiated in 1771 by Monge, both in geometric and analytic form, was briefly reprised by Euler in a purely analytic manner. The acclaimed Swiss mathematician introduced the coordinates (x, y, z) of a space curve parameterized as functions of arc length s. Euler unveiled equations for the osculating plane and the geodesics of a surface. Fourier, caught up in Monge's enthusiasm for space curves, established the relationship between the curvature and the torsion of a space curve. Fourier was the first to introduce the precise formula for the torsion, a concept that Monge hinted at in his initial study of 1771. It was not until 1806 when an enterprising young engineering student of Monge, Michel Ange Lancret (1774–1807), continued the study of space curves following Monge's exact methods. Lancret defined the curvature and torsion of a space curve as infinitesimal angles of rotation of the normal and osculating planes. He determined an explicit formula for the torsion of a curve in space in his "Mémoire sur les courbes à double courbure" (1806) and credited Fourier for his results.[49] In the spirit of his teacher, Lancret also explored problems for finding the evolutes of a curve.

---

[49]Coolidge, 1940, p.323 and p. 436.

3.3 Monge's "Two Curvatures of a Curved Surface"

A characteristic feature of Monge's mathematical approach to differential geometry was his purposeful use of geometric reasoning supported only by those analytical tools absolutely deemed necessary. Monge's discussion of the two curvatures of a curved surface provides a clear example of this attribute of his work and bears some profound results.

In Chapter XV, "Des deux courbures d'une surface courbe," of *Application de l'Analyse a la Géométrie*, Monge describes the concept of principal curvature. For plane curves, he explains that the center of the osculating circle may be thought of as the intersection of two consecutive normals to the curve. However, with regard to surfaces, the normal at any point will not meet the normal at a consecutive (nearby) point taken arbitrarily. If the consecutive point is taken in one of the principal directions (i.e., a direction in which the curvature is an extremum), then the two consecutive normals will intersect (i.e., the two normals will lie in the same plane), and their resultant length will be the corresponding principal radius of curvature. In other words, at any point on a surface there are two directions, at right angles to each other, such that the normal at a consecutive point taken on either line of intersection crosses the original normal. These directions are the two principal directions at that point.

Monge establishes these ideas mathematically beginning with the equation of a sphere.[50]

$$(x - x')^2 + (y - y')^2 + (z - z')^2 = R^2 \qquad (10)$$

where x', y', and z' are constants. The partial derivatives of (10) with respect to x, then y, are found successively to produce the

---

[50]Struik, 1950, pp. 414-415.

equations of two normal planes to the surface passing through the point (x', y', z'), where

$$p = \frac{dz}{dx} \quad \text{and} \quad q = \frac{dz}{dy}$$

are the partial derivatives. The two normal planes to the surface are given by

$$(x - x') + (z - z')p = 0 \qquad (11)$$
$$(y - y') + (z - z')q = 0 \qquad (12)$$

Since these planes are perpendicular to the xz-plane and the yz-plane, respectively, they are the equations of the projections of the normal to the surface. Monge sets forth the partial derivatives of second order (*différences partielles du second ordre*) r, s, and t in the equations:

$$dr = p\,dx + q\,dy$$
$$dp = r\,dx + s\,dy$$
$$dq = s\,dx + t\,dy$$

Differentiating (11) with respect to x and (12) with respect to y, where x', y' and z' are constants, we have

$$dx + p^2\,dx + pq\,dy + (z - z')(r\,dx + s\,dy) = 0$$
$$dy + pq\,dx + q^2\,dy + (z - z')(s\,dx + t\,dy) = 0$$

Geometrically, Monge explains these five equations in the following manner:

> If, from the first point of the surface under consideration, we pass in a certain direction to another point at an infinitesimal distance, then the five quantities x,

y, *z*, *p*, *q* will increase by their respective differentials *dx*, *dy*, *dz*, *dp*, *dq*. ... The value of the quantity $\frac{dy}{dx}$ will determine, in the plane of the *x*, *y*, the projection of the direction along which we pass from the first point to the second. ... This being established, if we conceive through the second point a new normal to the curved surface, and if this normal is in the same plane as the first and hence intersects it somewhere at a point, then this point of intersection will be that of the first normal for which the three coordinates x', y', z' do not change when *x* and *y* change their values.[51]

When $\frac{dy}{dx}$ then $(z - z')$ are eliminated, in order, from the two above equations we obtain:

$$(z - z')^2(rt - s^2) + \\ (z - z')\left[(1 + q^2)r - 2pqs + (1 + p^2)t\right] + \\ 1 + p^2 + q^2 = 0 \tag{13}$$

$$\frac{dy^2}{dx^2}\left[(1 + q^2)s - pqt\right] + \\ \frac{dy}{dx}\left[(1 + q^2)r - (1 + p^2)t\right] - \\ (1 + p^2)s + pqr = 0 \tag{14}$$

The point of intersection (x', y', z') of the two consecutive normals can be computed from equations (11), (12), and (13). From (14), $\frac{dy}{dx}$ can be computed and its value will specify the direction in which we must travel from the first point on the surface to the second. This will ensure that the new normal will be in the same plane as the first normal, and that they will have the point (x', y', z') in common.[52]

---

[51] *Ibid.*
[52] *Ibid.*

Equation (14) provides two values of $\frac{dy}{dx}$, since it is of second degree. Thus, from each normal to a point on the surface, we can move along the surface in two different directions to another nearby point on the surface. The normal at the new point will be in the same plane as the first normal. "The two directions in question have between them a very remarkable property, and that is that they are at right angles."[53] This means that tangent plane to the surface at the point (x, y, z) will be parallel to the xy-plane, which will make $p = 0$ and $q = 0$. We can rewrite (14) as

$$\frac{dy^2}{dx^2} + \frac{dy}{dx}\left(\frac{r-t}{s}\right) - 1 = 0.$$

Let m and m' be the two values of $\frac{dy}{dx}$ (i.e., the two solutions to the above equation), then

$$mm' + 1 = 0.$$

Monge concludes this section with the following explanation:

> Hence, every curved surface has at every one of its points two curvatures whose directions are in two normal planes perpendicular to one another, and whose centers are on the same normal.

> The three quantities x, y, z, being the coordinates of the point of the surface, and the other three x', y', z' being coordinates of the center of curvature, it is evident that the distance of these two points, that is to say, the value of the radius of curvature, is nothing else but the quantity R of equation (10). Hence, if between the four equations (10), (11), (12), and (13) the three quantities $x - x'$, $y - y'$, $z - z'$ are eliminated, then we shall have an equation of the second degree which will give in p,

---

[53] Monge, 1807, Chapter XV Section II, and Struik, 1950, p. 415.

q, r, s, t, the two values of R, hence those of the two radii of curvature. [54]

3.4 Maple Exposition of Normal and Principal Curvature

The concept of surface curvature is more intricate to describe than the curvature of a plane curve or a space curve. To detail how a surface bends at a point we will need to measure the curvature of the curves lying on the surface. We can define these curves by slicing the surface in a particular way and studying the resulting curves in order to gain insight into the nature of the curvature of the surface. From knowing how a tangent line is used to approximate a curve near a point (using linear approximation), we can use a tangent plane to approximate a surface near a point on the surface. Suppose p is a point on a surface S and $N(p)$ is a unit normal vector at p. Let **t** be a unit tangent vector in the tangent plane $T_p(S)$. Then the plane spanned by the vectors $N(p)$ and **t** will be the vertical plane through p. The intersection of this plane with the surface S is a curve on the surface through p called a *normal section* of S at p. The frames from a Maple animation shown in Figures 3.25, 3.26, and 3.27 portray a series of normal sections which are curves on the parametrized saddle surface

$$f(u,v) = \left(u, v, -u^2 + v^2\right).$$

The curvature of each normal section is called the *normal curvature* of S at point p. As the vertical plane rotates around point p, the radius of curvature will change. At each point p, there are two principal radii; for some particular normal section, it will reach a minimum value R_1 and for some other normal section there will

---

[54]Struik, 1950, pp. 416-417.

be a maximum value R_2 Thus, we call κ_1 and κ_2 the *principal curvatures*. Geometrically, we have the following formulas:

$$\kappa_1 = \frac{1}{R_1} \quad \text{and} \quad \kappa_2 = \frac{1}{R_2}$$

The directions of the unit tangent vectors **t** to the two normal sections for which the curvature is a minimum or maximum are called the *principal directions* of the surface S at point p.

It was Euler who first discovered that the principal directions at a point p on S are always perpendicular.[55] Euler's work in the subject also demonstrated that the principal curvatures and the angle that the normal section makes with the principal directions completely determine the curvature of the normal section.

The surface in the following series of Maple plots is a saddle surface and the point p at the origin is a hyperbolic point. The center of curvature for each normal section does not remain on one side of the surface. If the normal section passes through the left and right sides of the saddle, i.e., the high parts, the center of curvature will be located on the line through the normal above the point p since the curve of intersection is concave up as shown in Figure 3.25. In this case, the normal section is bending in the same direction as the surface normal. That is, in the direction of the unit tangent vector at p the surface is bending toward the surface normal. This means that the sign of the curvature of the normal section is positive, i.e., $\kappa(p) > 0$. If we look at all of the normal sections with centers of curvature above the point p there will be one with the largest, or most positive, curvature for the saddle surface that we call κ_1.

The Maple animation code for the normal sections in Figures 3.25, 3.26, and 3.27 is given in Listing 3.16.

---

[55]Struik, 1950, p. 416.

```
f := (u, v) -> [u, v, -u^2+v^2]

surface := plot3d(f(u, v),
    u = -2 .. 2, v = -2 .. 2,
    style = patchnogrid, shading = xyz,
    lightmodel = light2):

surfacenormal := arrow([0, 0, 0], [0, 0, 6],
    .1, .3, .2, color = black):

numframes := 16:

normalplane := animate3d(
    [r*cos(theta), r*sin(theta), s],
    r = -2 .. 2, s = -3 .. 5,
    theta = 0 .. (numframes-1)/numframes*2*Pi,
    frames = numframes,
    color = brown, thickness = 2, style = wireframe):

display({surface, normalplane, surfacenormal})
```

Listing 3.16: Animation of normal sections on saddle surface

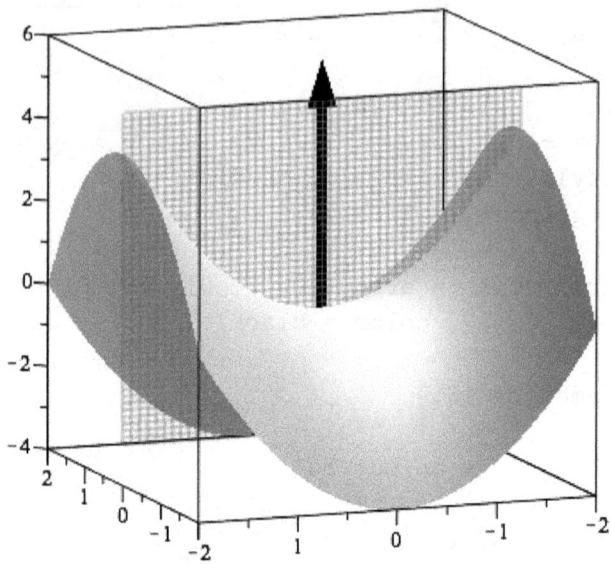

Figure 3.25: Normal section of saddle surface
with $\kappa_1 > 0$

The situation shown in Figure 3.26 illustrates how the curvature will decrease as the vertical plane is rotated out of the position given in in Figure 3.25. As the length of the radius of curvature increases, the curvature will be zero, i.e., $\kappa(p) = 0$. This occurs when the curve of intersection becomes a straight line as the normal plane moves in the direction where the surface intersects its tangent plane at p.

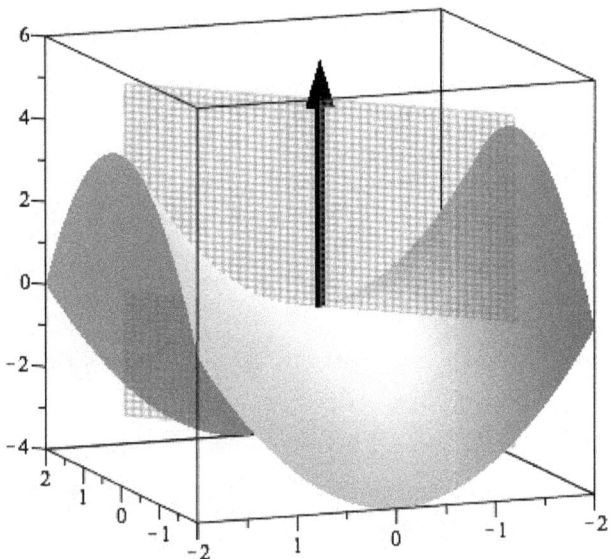

Figure 3.26: Normal section of saddle surface with $\kappa_1 = \kappa_2 = 0$

If the normal section passes through the front and back, i.e., the lower sections, of the saddle as shown in Figure 3.26, the center of curvature will lie below the point p. In this case, the center of curvature will move upward on that line approaching p with each rotation out of this position. Therefore, the center of curvature will be increasing as the radius of curvature is decreasing. Ultimately, the curvature will reach its most negative value, call it κ_2 of all the normal sections of the saddle surface with center of curvature below p. Here we see the normal section bending in the opposite direction from the surface normal. As the surface is bending away from the normal in the direction of the tangent at p, the sign of the curvature of the normal section is negative, i.e., $\kappa(p) < 0$. It is of importance to note here that the principal curvatures can be changed by the bending of the surface, thus, they are called *extrinsic quantities*.

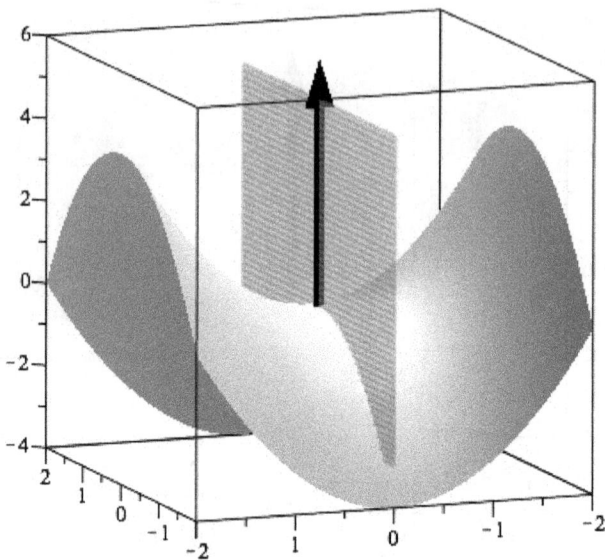

Figure 3.27: Normal section of saddle surface with $\kappa_2 < 0$

3.5 Lines of Curvature

In Chapter XVI entitled, "Des lignes de courbure de la surface de l'ellipsoide," of *Application de l'Analyse à la Géométrie* , Monge shows that the lines of curvature on a surface possess the property in which the normals along them generate developable surfaces. Lines of curvature drawn on a surface may be defined equivalently in the following ways:

- they are tangent at each point to one of the principal directions;

- they have zero geodesic torsion; and

- the associated normal surface is developable.[56]

---

[56]Sakarovitch, 2009, p.1294.

Monge demonstrated that along two families of lines of curvature, the normals to the lines of curvature form two families of developable surfaces (or normal surfaces) to the lines of curvature. In particular, Monge studied the lines of curvature of the ellipsoid in Figure 3.28. He concluded that the lines of curvature to this particular surface are a ruled (but not developable) family of curves, of which their projections are a family of ellipses or a family of hyperbolas.[57] A physical and practical application of Monge's work on lines of curvature relates to architecture and the construction of ellipsoid vaults. These types of extended arches are built using spherical vaults or ellipsoids of revolution with horizontal layers. They are used to create large open areas in a building and are common in classical and Baroque architecture and provide a prime example of how Monge melded geometry with construction.[58]

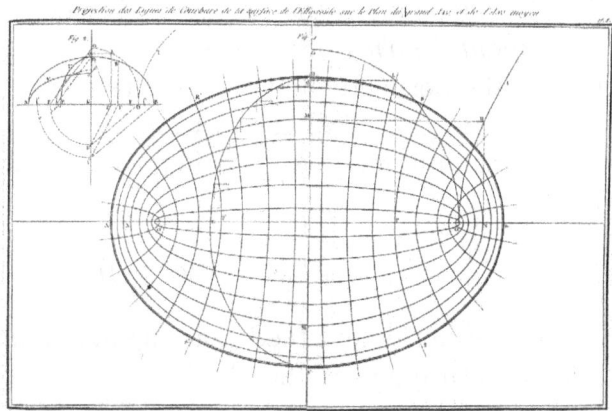

Figure 3.28: Lines of curvature on an ellipsoid from *Application de l'Analyse à la Géométrie* (1807)

---

[57]Sakarovitch, 2009, p.1297.
[58]*Ibid.*

3.6 Historical Perspective on Curvature of Surfaces

Euler was the first to study curvature of surfaces. His paper, "Recherches sur la courbure des surfaces" (1760), was the premiere publication on surface theory.[59] Clairaut had done some work thirty years earlier on geodesics, finding the tangent plane at a point on a surface, and on surfaces of revolution. However, Euler's method for finding the curvature of a surface at a given point was a giant leap forward for the beginnings of differential geometry.

Euler showed that the curvature of a normal section is given by the equation

$$k_\phi = L + M\cos(2\phi) + N\sin(2\phi)$$

where L, M, and N depend on the partial derivatives (coefficients of the *first fundamental form*) of the surface $z = f(x,y)$ at p and ϕ is the angle that the perpendicular to the surface makes with the principal plane, i.e., the normal plane in which the curvature takes on its maximum or minimum value. Differentiating with respect to ϕ we have

$$-2M\sin(2\phi) + 2N\cos(2\phi) = 0.$$

From this equation, Euler determined that the maximum and minimum values of curvature occur when

$$\tan(2\phi) = \frac{N}{M}.$$

Using the identity $\tan(2\phi) = \tan(2\phi + 2\pi)$, Euler gave justification that if the maximum value of curvature k_1 happens at a particular value of ϕ, then the minimum value of curvature k_2 must occur at $\phi + 2\pi$. The mathematical conclusion is *Euler's theorem on*

---

[59]Struik, 1933, p. 103.

curvature of surfaces which states that the curvature of any normal section at an angle of ϕ to the principal plane is given by

$$k = \frac{k_1 + k_2}{2} - \frac{(k_1 - k_2)\cos(2\phi)}{2}.$$

The formula simplifies to

$$k = k_1 \cos^2\theta + k_2 \sin^2\theta,$$

where k is the normal curvature to the surface at point p, θ is the angle between a direction vector at p and the first principal direction, and k_1, k_2 are the principal curvatures.

One of Monge's most able students at the École Royale du Génie at Mézières was Jean-Baptiste Meusnier (1754–1793). Meusnier achieved Euler's results on principal curvatures in a new and elegant way. He presented a paper "Memoire sur la courbure des surfaces" to the Paris Académie des Sciences in 1776.[60]

Both Dupin and Olinde Rodrigues (1794–1851) continued Monge's work on lines of curvature, simplifying his results. Dupin set up three families of surfaces which intersect orthogonally. He proved that the curve of intersection of any two surfaces in this triply orthogonal system is a line of curvature on each surface, this result is known as *Dupin's theorem*. Rodrigues clarified Monge's findings on the lines of curvature. He also found a formula for the curvature of a surface at a point, ahead of Gauss, via a spherical mapping of the surface and by studying the ratios of areas of the corresponding surfaces. Rodrigues showed that the total curvature was equal to the product of the two principal curvatures. Euler's work in surface theory continued in a new direction during the decade of 1770. His investigations involved surfaces that could be formed by bending plane regions, i.e., developable surfaces. Euler was the first to represent a point on a surface as a function of two

---

[60]Sakarovitch, 2009, p.1294.

variables, $X(t,u)$. However, it was Carl Friedrich Gauss (1777-1855) who used this parametrization throughout his examination into the theory of surfaces.

Monge pursued the work Euler began in the theory of developable surfaces, formalizing Euler's findings. He gave shape to the theory in his paper of 1785 entitled, "Mémoire sur les développées, les rayons de courbure, et les différens genres d'inflexions des courbes à double courbure." It was Monge who outlined the essential differences between ruled and developable surface and determined that developable surfaces must satisfy the differential equation given by $rt - s^2 = 0$.

Gauss shed new light on Euler's results on the curvature of surfaces. He knew the importance and essential elements of intrinsic surface theory and worked to developed the equations and forms of curvature that comprise differential geometry.[61] In 1828, Gauss wrote, "Disquisitiones generales circa superficies curvas," enormously influencing the trajectory of differential geometry in the nineteenth century. His skillfully written and perfectly refined paper was brimming with original ideas and explorations into the intrinsic properties of surfaces. Through this important paper, Gauss primarily gave differential geometry its present form. He generated new and important problems for geometers to work on for many decades to come.[62] Gauss formulated the outcome that the curvature K is a function of the coefficients of the first fundamental form and its derivatives, that is,

$$K = k_1 k_2$$
$$= \frac{eg - f^2}{EG - F^2}.$$

He formulated one of the most renowned theorems of the nineteenth century, his "great theorem," *Theorema Egregium* (1828)

---

[61]Sakarovitch, 2009, p. 1294.
[62]Kolmogorov, 1996, p. 7.

which proved that each of the principal curvatures k_1, k_2 individually depend on the extrinsic properties of the surface (i.e., how the surface bends in the coordinate space in which it is defined). For example, the slope of a tangent line is an extrinsic property. Whereas, the product of the two principal curvatures depends only on the intrinsic geometry of the surface. In other words, the Gaussian curvature of a surface at a point given by the product of the principal curvatures $K = k_1 k_2$ can be ascertained from the characteristics of the surface itself without considering the larger space. For instance, the length of a curve on a surface is an intrinsic property of the surface.

The Gaussian curvature reveals certain properties about the surface S at the point p. In particular, it is possible to distinguish between four kinds of points p at which S is cap-shaped (or elliptic), saddle shaped (or hyperbolic), planar, or parabolic (an intermediate state between elliptic and hyperbolic) in shape:

1. if $k_1 k_2 > 0$, then p is an elliptic point

2. if $k_1 k_2 < 0$, then p is a hyperbolic point

3. if $k_1 k_2 = 0$, that is, if one of the principal curvatures is zero, then p is a parabolic point

4. if $k_1 k_2 = 0$ and both of the principal curvatures are zero, then p is a planar point.

Gauss also found that if a curved surface is developed (i.e., unrolled flat or isometrically mapped) onto another surface, then the measure of the curvature at each point will remain invariant.

In the last half of the nineteenth century, there was tremendous growth and advancement in differential geometry from Bernhard Riemann (1826–1866) to Albert Einstein (1879–1955) through the varied schools of geometers in between and beyond. Concepts of differential geometry in three dimensions were extended to n-dimensions by Riemann and he charted theories on curvature of surfaces in n-dimensions. Riemann's ideas of curvature in higher dimensions were necessary for Einstein to complete his general theory of relativity. As the field of differential geometry expanded, it became more of an abstract study and one of the most fascinating areas in contemporary mathematics. The subject grew a long way from its origins in the practical problems posed by Monge, who is often regarded as the "father of differential geometry".

3.7 Maple Exposition of Ruled and Developable Surfaces

Monge's theory of ruled and developable surfaces in Chapters XI through XIII of *Application de l'Analyse à la Géométrie* was analytically developed from the geometrical topics he had first considered in *Géométrie Descriptive*. Monge's ideas on ruled and developable surfaces grew from the empirical problems of the time. He was motivated by architects, stone cutters, and sculptors who wished to design curved surfaces by using a straight edge along a stationary line (i.e., the z-axis). Monge was among the first mathematicians to investigate ruled surfaces. In each chapter, Monge classifies problems based on their relation to partial differential equations of the first, second, and third order. Developables and ruled surfaces with generators parallel to a given plane are obtained from second order differential equations and ruled surfaces are obtained from partial differential equations of the third order.[63]

A *ruled surface* is a surface swept out by a moving straight line in $\mathbb{R}^3$ such that the surface is composed entirely of straight lines, called generators or rulings. A ruled surface has parametrization

$$X(u,v) = \beta(u) + v\delta(u).$$

---

[63]Struik, 1933, p.106.

The surface is made up of lines radiating from a curve $\beta(u)$ moving in the direction of $\delta(u)$. The base curve $\beta(u)$ is called the *directrix* and it is intersected by each ruling of the surface. Thus, a ruling is a line with $\delta(u)$ as direction vector.

3.7.1 Cylinders, Cones, Hyperboloids of Revolution

Some examples of ruled surfaces are a cylinder

$$X(u,v) = (a\cos u, a\sin u, v),$$

a cone

$$X(u,v) = (v\cos u, v\sin u, v)$$

and a hyperboloid of revolution

$$X(u,v) = (\cos u - v\sin u, \sin u + v\cos u, v),$$

The Maple plots for these three surfaces, with rulings clearly visible, are displayed in Figures 3.29, 3.30, and 3.31.

```
ruledcyl := (u, v) -> [cos(u), sin(u), v]

plot3d(ruledcyl(u, v),
    u = 0 .. 2*Pi, v = 0 .. 2*Pi,
    axes = none, shading = xy, lightmodel = light1,
    style = surfacewireframe,
    scaling = unconstrained);

ruledcone := (u, v) ->[v*cos(u), v*sin(u), v]

plot3d(ruledcone(u, v),
    u = 0 .. 2*Pi, v = -3 .. 3,
    axes = none, shading = xy, lightmodel = light1,
    style = surfacewireframe,
    scaling = unconstrained,
    orientation = [47, 76, 0]);

x := u -> cos(u):
y := u -> sin(u):
z := u -> 0:

w1 := u -> -sin(u):
w2 := u -> cos(u):
w3 := u -> 1:

ruledhyperboloid :=
(u,v) -> [x(u)+v*w1(u), y(u)+v*w2(u), z(u)+v*w3(u)]:

plot3d(ruledhyperboloid(u, v),
    u = 0 .. 2*Pi, v = -3 .. 3,
    axes = none, shading = xy, lightmodel = light1,
    style = surfacewireframe,
    orientation = [34, 79, -4]);
```

Listing 3.17: Ruled surfaces: cylinder, cone and hyperboloid of revolution

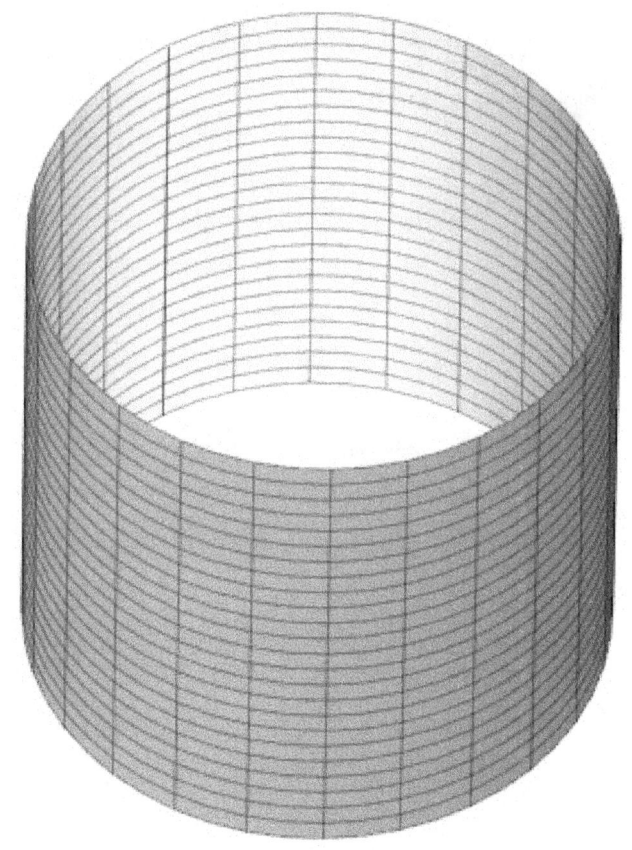

Figure 3.29: Ruled cylinder

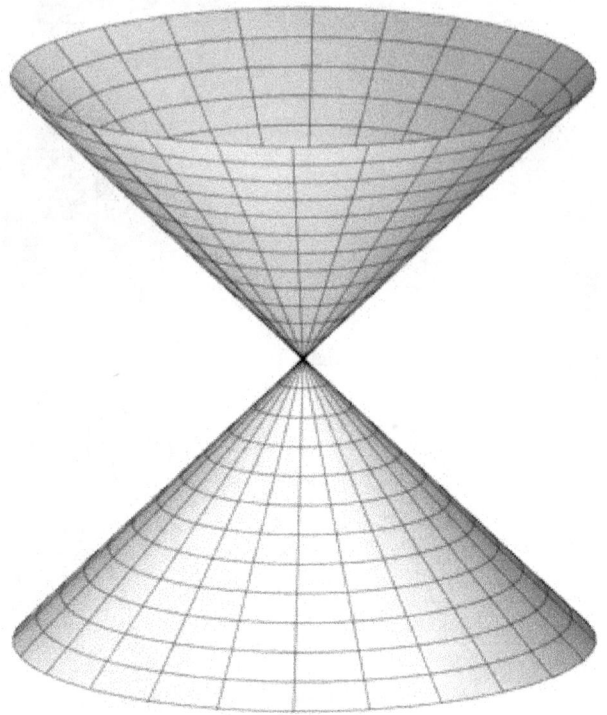

Figure 3.30: Ruled cone

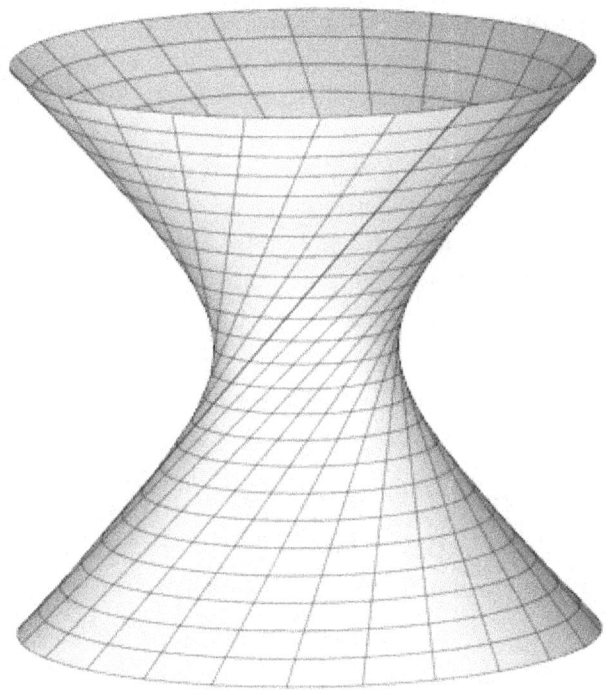

Figure 3.31: Hyperboloid of revolution

Monge classified ruled surfaces as either *developable* or *warped*. He asserted that a developable surface is a ruled surface that can be unrolled onto a plane and will lie flat, whereas, a warped surface will not. Thus, the rulings of a developable surface are straight, parallel lines. A warped surface will be bent or twisted when unrolled and its rulings are skew lines. The cylinder and the cone in Figures 3.29 and 3.30, respectively, are developable surfaces. The hyperboloid of revolution in Figure 3.31 is an example of a warped surface generated by revolving one skew line about another skew line. Ruled and developable surfaces were first investigated in descriptive geometry since it was possible for these types of surfaces to be constructed by purely geometrical means (e.g., the surfaces produced from Monge's silk thread models and Olivier's string models).

Figure 3.32 provides another view of the hyperboloid of revolution with two of its rulings highlighted. The rulings of the surface are the straight lines

$$L(v) = \beta(u) + v\delta(u),$$

where each value of u generates a fixed line as v varies. From this graphic it is possible to see that the rulings are, in fact, skew lines. The Maple code use to generate each specific ruling is shown in Listing 3.18.

Thus far, we have seen three elementary examples of ruled surfaces. In the spirit of Monge and his knack for describing surfaces in three dimensions, Maple will be used to bring three unique ruled surfaces to life. Figures 3.33 and 3.34 display a ruled cylinder and ruled cone with a base curve (directrix) in the shape of a figure eight. The surface remains ruled even when the base curve is changed.

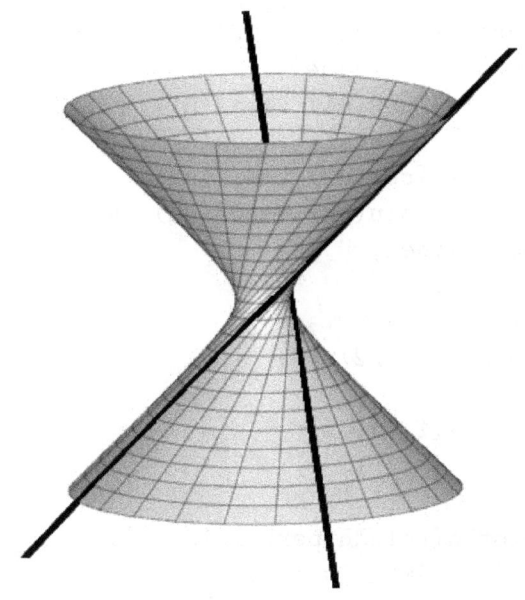

Figure 3.32: Ruled hyperboloid of revolution with two generators highlighted

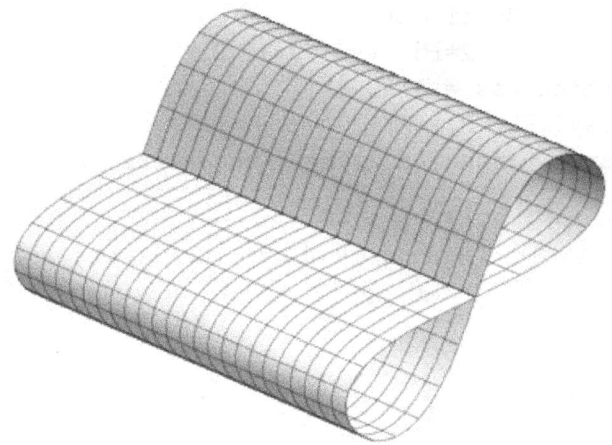

Figure 3.33: Ruled cylinder with figure eight base curve

```
generators := proc(u1, u2)
    local line1, line2;
    line1 := [cos(u1)-v*sin(u1),
              sin(u1)+v*cos(u1), v];
    line2 := [cos(u2)-v*sin(u2),
              sin(u2)+v*cos(u2), v];
    line1, line2;
 end:

generators(0, (1/2)*Pi)

line1 := v->[1, v, v]:
line2 := v->[-v, 1, v]:

P1 := plot3d(ruledhyperboloid(u, v),
    u = 0 .. 2*Pi, v = -7 .. 7,
    color = tan, lightmodel = light1,
    style = surfacewireframe):
P2 := plot3d(line1(v),
    u = 0 .. 2*Pi,
    v = -9 .. 9, thickness = 4):
P3 := plot3d(line2(v),
    u = 0 .. 2*Pi, v = -9 .. 9,
    thickness = 4):
display({P1, P2, P3},
    axes = none, scaling = unconstrained);
```

Listing 3.18: Generators of hyperboloid of revolution

```
p1 := cos(u):
p2 := sin(u)*cos(u):
p3 := 0:
q1 := 0:
q2 := 1:
q3 := 1:

cylinder8 := (u, v) -> [p1+v*q1, p2+v*q2, p3+v*q3]

plot3d(cylinder8(u, v),
    u = 0 .. 2*Pi, v = 0 .. 4,
    axes = none, shading = xy, lightmodel = light2,
    style = surfacewireframe,
    scaling = unconstrained,
    orientation = [44, 9, 17]);

p1 := 1:
p2 := 1:
p3 := 0:
q1 := cos(u):
q2 := sin(u)*cos(u):
q3 := 2:

cone8 := (u, v)-> [p1+v*q1, p2+v*q2, p3+v*q3]

plot3d(cone8(u, v),
    u = 0 .. 2*Pi, v = -2 .. 2,
    axes = none, shading = xy,
    lightmodel = light1, style = surfacewireframe,
    scaling = constrained,
    orientation = [-171, 20, -48]);
```

Listing 3.19: Modified ruled cylinder and cone

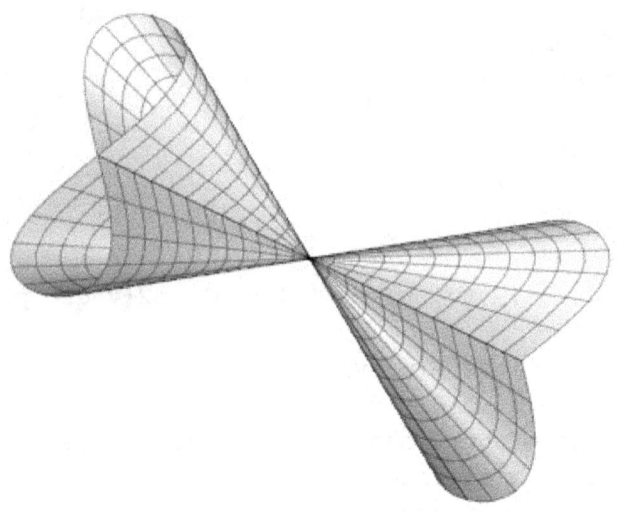

Figure 3.34: Ruled cone with figure eight base curve

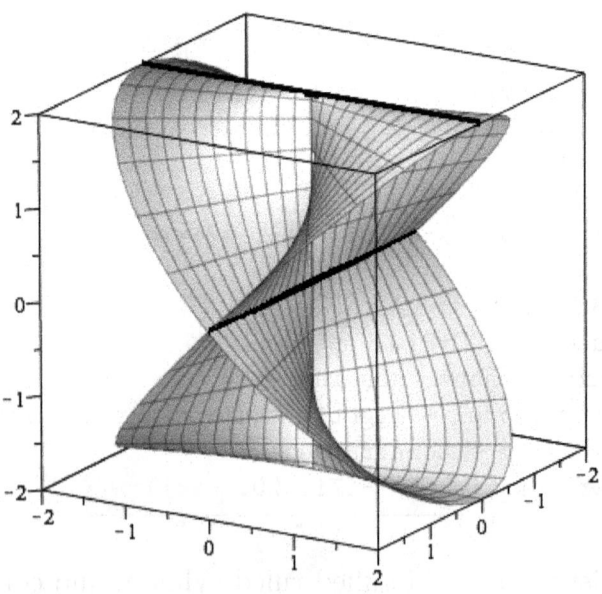

Figure 3.35: Right conoid with two rulings highlighted

Lastly, we observe that the generators of the intriguing right conoid ruled surface in Figure 3.35 are parallel to the xy-plane and pass through the z-axis. A right conoid is a ruled surface that is swept out by a family of horizontal lines that are perpendicular to a fixed vertical line.

```
generators := proc(u1, u2)
    local line1, line2;
    line1 := [v*cos(u1), v*sin(u1), 2*sin(u1)];
    line2 := [v*cos(u2), v*sin(u2), 2*sin(u2)];
    line1, line2;
 end:

generators(0, (1/2)*Pi)

rtconoid := (u, v)->[v*cos(u), v*sin(u), 2*sin(u)]

line1 := v->[v, 0, 0]:
line2 := v->[0, v, 2]:

P1 := plot3d(rtconoid(u, v),
    u = -Pi .. Pi, v = -2 .. 2,
    axes = boxed, shading = xyz, lightmodel = light1,
    style = surfacewireframe):
P2 := plot3d(line1(v),
    u = 0 .. 2*Pi, v = -2 .. 2,
    color = brown, thickness = 4):
P3 := plot3d(line2(v),
    u = 0 .. 2*Pi, v = -2 .. 2,
    color = brown, thickness = 4):
display({P1, P2, P3},
    axes = boxed, scaling = constrained);
```

Listing 3.20: Generators of right conoid

3.7.2 Developable Surfaces and Gaussian Curvature

Geometrically, we can build developable surfaces by: (a) rolling a piece of paper into a cylinder, or (b) bending a piece of paper in the shape of a circular sector into a cone, or (c) cutting a circle out of a piece of paper and then twisting the portion that remains to form a circular helix. The surface constructed in (c) is called a *tangent developable surface* in which the tangent lines to the circle become the generating lines (rulings) of the surface. The cylinder, cone and tangent developable surface all share a common characteristic; in order to unroll each of these surfaces onto the plane, the tangent plane must be constant along each generating line. This means that the tangent planes must only depend upon one parameter. Thus, we can consider these surfaces as Monge did, as the envelope of a one-parameter family of planes. Monge sought to find a necessary and sufficient condition that would distinguish ruled surfaces from developable surfaces. His work in setting up the differential relation which states the condition that all developable surfaces must satisfy is included in Chapter XII "Des Surfaces Développables" of *Application de l'Analyse à la Géométrie*. The following discussion outlines Monge's analysis using modern notation and analytical procedures.

Given a ruled surface

$$z = f(x, y).$$

Let

$$p = \frac{\partial z}{\partial x} \quad \text{and} \quad q = \frac{\partial z}{\partial y}.$$

The tangent plane to the surface at the point (a, b, c) is

$$(z - c) = (x - a)p + (y - b)q.$$

As the point (a, b, c) varies along one of the rulings of the surface, the tangent plane remains constant along a ruling, so the equation of the plane will not change. This implies that p and q do not vary, thus $dp = dq = 0$. Monge defines

$$dp = r\,dx + s\,dy \quad \text{and} \quad dq = s\,dx + t\,dy$$

with

$$r = \frac{\partial p}{\partial x}, s = \frac{\partial p}{\partial y} = \frac{\partial q}{\partial x}, t = \frac{\partial q}{\partial y}$$

and sets

$$dp = r\,dx + s\,dy = 0 \quad \text{and} \quad dq = s\,dx + t\,dy = 0$$

Solving the system for $\frac{dy}{dx}$, the direction in which the tangent is zero, using the Jacobian determinant we have

$$\frac{\partial(p,q)}{\partial(x,y)} = \begin{vmatrix} \frac{\partial p}{\partial x} & \frac{\partial p}{\partial y} \\ \frac{\partial q}{\partial x} & \frac{\partial q}{\partial y} \end{vmatrix} = 0$$

thus,

$$\frac{\partial(p,q)}{\partial(x,y)} = rt - s^2 = 0.$$

Monge concluded that all surfaces $f(x,y)$ which satisfy the equation $rt - s^2 = 0$ are developable surfaces. The components r, t, s of Monge's expression $rt - s^2$ are equivalent to e, g, f, respectively, in the *second fundamental form* given a regular parameterized surface $X(u,v)$ in $\mathbb{R}^3$:

$$\text{II} = e\,du^2 + 2f\,du\,dv + g\,dv^2.$$

Note that Monge's expression $rt - s^2$ is the numerator of the *Gaussian curvature*[64], or total curvature, of a surface

$$K = \frac{eg - f^2}{EG - F^2}.$$

---

[64] $E, F,$ and G are the coefficients of the first fundamental form and $e, f,$ and g are the coefficients of the second fundamental form.

Hence, the necessary and sufficient condition for a surface to be developable is that $K = 0$ at every point on the surface. The formulas for the coefficients E, F, G and e, f, g of the first and second fundamental forms, respectively, are given in the examples that follow. They will be utilized to compute the Gaussian curvature via Maple of surfaces with parametrizations $X(u, v)$.

Recall that a *tangent developable* surface is a ruled surface generated by the tangent lines to a space curve. This makes the tangents of the space curve the rulings of the surface. The tangent developable consists of two sheets meeting along a space curve in a sharp edge or cusp, called the *edge of regression*. Given a space curve $\beta(u)$, the parametrization of its tangent developable surface is

$$X(u, v) = \beta(u) + v\beta'(u).$$

The edge of regression of the tangent developable surface in Figure 3.36 is a helix with parametrization

$$\beta(u) = \left(\sin(2u), \cos(2u), \frac{u}{2}\right).$$

The Maple code produces each half-tangent in contrasting colors to emphasize the two sheets of the surface and its sharp edge. The tangent developable surface to Viviani's curve with edge of regression

$$\beta(u) = \left(1 + \cos u, \sin u, 2\sin\left(\frac{u}{2}\right)\right)$$

is given in Figure 3.37.

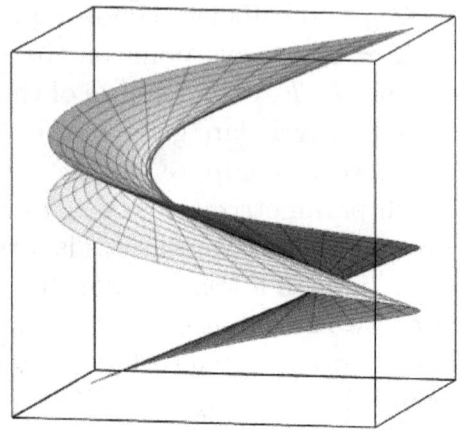

Figure 3.36: Tangent developable surface of a helix

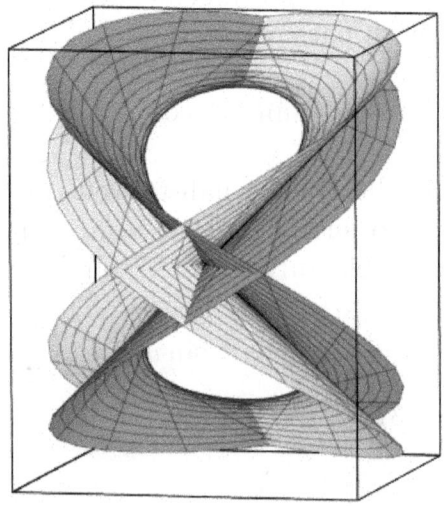

Figure 3.37: Tangent developable surface to Viviani's curve

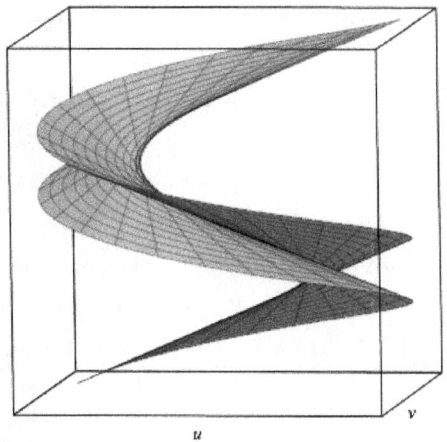

Figure 3.38: Tangent developable surface to a helix shaded by its Gaussian curvature: $K = 0$

Up to this point, we have used Maple to construct ruled surfaces that are already known to be developable surfaces. It would be useful to devise a test which could be used to verify whether or not a ruled surface is developable. Recall that the necessary and sufficient condition for a surface to be developable is $K = 0$. Therefore, we can use Maple to color a surface by its Gaussian curvature. Figure 3.38 shows the tangent developable surface to the helix under consideration. The surface has parametrization

$$X(u, v) = \left(\sin(2u) + 2v\cos(2u), \cos(2u) - 2v\sin(2u), \frac{u+v}{2} \right).$$

Listings 3.22 and 3.23 outline how to compute and color a surface by its Gaussian curvature, K. The Maple output displays the entire surface in one color, indicating that K is constant throughout the surface. Specifically, since this surface is known to be developable, the Maple output shows that a monochromatic plot must mean $K = 0$.[65]

In Figure 3.39, the hyperboloid of revolution with parametrization
$$X(u, v) = (\cos u - v \sin u, \sin u + v \cos u, v)$$
is shaded by K using Maple. The plot confirms our earlier work: this surface is not developable. The Gaussian curvature of the hyperboloid surface is not constant and most certainly not zero everywhere. Since the rulings of the hyperboloid of revolution are skew, the tangent plane will vary as the point of tangency moves along one of the rulings. This is illustrated by the color variation of the curvature values around the waist of the surface in comparison with the color representing the curvature values at the top and bottom of the surface. The following formulas were applied to compute K given $X(u,v)$:

$$N(u,v) = \frac{X_u \times X_v}{|X_u \times X_v|}$$
$$E = X_u \cdot X_u$$
$$F = X_u \cdot X_v$$
$$G = X_v \cdot X_v$$
$$e = X_{uu} \cdot N$$
$$f = X_{uv} \cdot N$$
$$g = X_{vv} \cdot N$$
$$K = \frac{eg - f^2}{EG - F^2}$$

where $N(u,v)$ is the unit normal to the surface.

---

[65] When the Gaussian curvature is computed for a surface and then used to color that surface, Maple 2015 assigns the color RED to indicate that $K = 0$. If a surface is shaded by its mean curvature H and $H = 0$, i.e., the surface is minimal, Maple 2015 renders the plot in RED as well.

```
tandev := proc(curve)
    local curveprime;
    curveprime := map(diff, curve, u);
    [v*curveprime[1]+curve[1],
    v*curveprime[2]+curve[2],
    v*curveprime[3]+curve[3]];
 end:

helix := <sin(2*u), cos(2*u), (1/2)*u>;
tandev(helix);

plot3d(tandev(helix),
    u = 0 .. Pi, v = -1 .. 1,
    color = COLOR(RGB, 1/4+v, 1/4-v, -(1/4)*v),
    style = patch, lightmodel = light2,
    style = surfacewireframe,
    scaling = unconstrained,
    axes = boxed);

Viviani := <1+cos(u), sin(u), 2*sin((1/2)*u)>;
tandev(Viviani);

plot3d(tandev(Viviani),
    u = 0 .. 4*Pi, v = -2 .. 2,
    color = COLOR(RGB, 1/2-sin(v), 1/2+sin(v), 0),
    style = patch, lightmodel = light1,
    style = surfacewireframe, scaling = constrained,
    axes = boxed);
```

Listing 3.21: Tangent developable surfaces

```
GaussK := proc(X)
    local Xu, Xv, E, F, G, n, N,
              Xuu, Xuv, Xvv, e, f, g;
    Xu := <diff(X[1], u),
           diff(X[2], u),
           diff(X[3], u)>;
    Xv := <diff(X[1], v),
           diff(X[2], v),
           diff(X[3], v)>;

    E := DotProduct(Xu, Xu);
    F := DotProduct(Xu, Xv);
    G := DotProduct(Xv, Xv);

    n := CrossProduct(Xu, Xv);
    N := Norm(n);

    Xuu := <diff(Xu[1], u),
            diff(Xu[2], u),
            diff(Xu[3], u)>;
    Xuv := <diff(Xu[1], v),
            diff(Xu[2], v),
            diff(Xu[3], v)>;
    Xvv := <diff(Xv[1], v),
            diff(Xv[2], v),
            diff(Xv[3], v)>;

    e := DotProduct(Xuu, n/N);
    f := DotProduct(Xuv, n/N);
    g := DotProduct(Xvv, n/N);

    simplify((e*g-f^2)/(E*G-F^2));
end:
```

Listing 3.22: Computation of Gaussian curvature, K

```
helix := <sin(2*u), cos(2*u), (1/2)*u>;
helixtd := tandev(helix);
GaussK(helixtd);

plot3d(helixtd,
    u = 0 .. Pi, v = -1 .. 1,
    color = GaussK(helixtd),
    style = surfacewireframe,
    lightmodel = light2, scaling = unconstrained,
    axes = boxed);

hyperboloid := <cos(u)-v*sin(u), sin(u)+v*cos(u), v>;
GaussK(hyperboloid);

plot3d(hyperboloid,
    u = 0 .. 2*Pi, v = -3 .. 3,
    color = GaussK(hyperboloid),
    style = surfacewireframe,
    lightmodel = light4, scaling = constrained,
    axes = none);
```

Listing 3.23: Coloring surfaces by Gaussian curvature, K

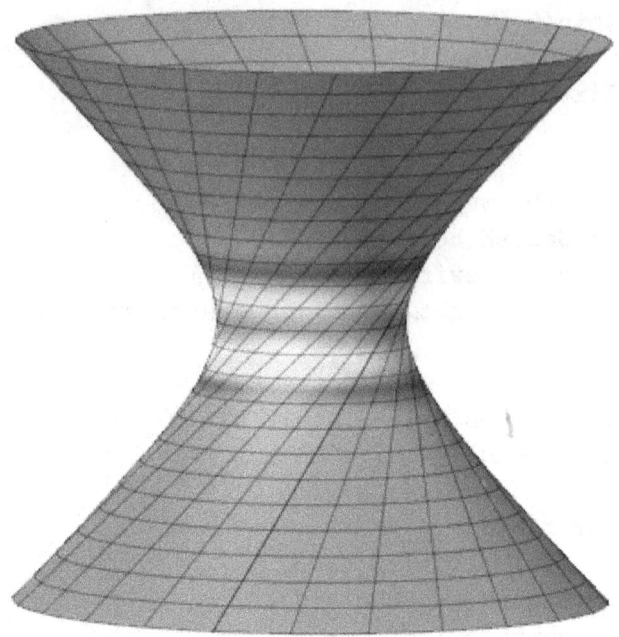

Figure 3.39: Hyperboloid of revolution shaded by Gaussian curvature: $K \neq 0$

3.7.3 Minimal Surfaces and Mean Curvature

The Gaussian curvature and mean curvature are entities of central importance when studying the geometry of surfaces. The Gaussian curvature is useful in determining if a surface can be developed into a plane. The *mean curvature* H for a regular surface in $\mathbb{R}^3$ with parametrization $X(u,v)$ is given by

$$H = \frac{k_1 + k_2}{2} = \frac{Ge + Eg - 2Ff}{2(EG - F^2)}$$

where E, F, G and e, f, g are the coefficients of the first and second fundamental forms, respectively.[66]

Coloring a surface according its Gaussian curvature or mean curvature can pinpoint sudden changes in the curvature, identify dimples or waves in the surface, and locate areas on the surface where the curvature may be higher or lower than neighboring areas. Surfaces for which $H = 0$ are called *minimal surfaces*. The focus of Meusnier's work was on minimal surfaces. Specifically, he examined characteristics of the catenoid and helicoid. Lagrange, in 1762, analytically derived the equation of minimal surfaces by employing general techniques of the calculus of variations. Lagrange did not provide a geometrical interpretation of his findings. He viewed "geometrical problems merely as illustrations for the application of the remarkable analytical methods that he developed."[67] Meusnier obtained Lagrange's equation geometrically and presented original corollaries resulting from his work.

In Section XX of *Application de l'Analyse à la Géométrie*, Monge discusses the situation in which the two principal curvatures at an arbitrary point on a surface have the same magnitude but are opposite in sign. Thus, the mean curvature is equal to zero at each point of the surface, which classifies it as a minimal surface. Monge specified an important property of minimal surfaces: "... if we surround part of the surface with a continuous closed contour,

---

[66]The mean curvature and Gaussian curvature formulas shown are computed using Calculus. We may also define these quantities using the shape operator from linear algebra. Given a point p on the surface S, the curvatures may be more formally defined as

$$K(p) = \det \begin{bmatrix} k_1(p) & 0 \\ 0 & k_2(p) \end{bmatrix} = k_1 k_2$$

$$H(p) = \text{trace} \begin{bmatrix} k_1(p) & 0 \\ 0 & k_2(p) \end{bmatrix} = \frac{k_1 + k_2}{2}.$$

[67]Fomenko, 1991, p. 24.

then of all surfaces passing through this contour, the area inside the contour is smallest."[68] The minimal surface formulas developed by Monge were specific to the helicoid and catenoid.

Another distinguished student of Monge, Siméon Denis Poisson (1781-1840), who authored more than 300 papers, further contributed to minimal surface theory through his inquiry into the theory of liquids and capillary effects. The essential properties of minimal surfaces were discovered when Joseph Plateau (1801–1883), a Belgian physicist, conducted experiments on soap films to find a surface of least area within a given boundary known as "Plateau's problem". Other examples of minimal surfaces are Scherk's surface, Catalan's surface, Enneper's surface, Henneberg surface, Bour's surface, and Costa's surface.[69]

The Maple procedure in Listing 3.24 provides a general method for computing the mean curvature of a regular surface in $\mathbb{R}^3$ given by $X(u,v)$. Figure 3.40 illustrates a hyperboloid of revolution shaded according to its mean curvature. The Maple output produced, upon executing the code in Listing 3.25, calculates the mean curvature as a function of the parameter v. We note that that mean curvature, H, of the hyperboloid is negative.

$$X(u,v) = (\cos u - v \sin u, \sin u + v \cos u, v)$$

and

$$H = -\frac{v^2}{(2v^2 + 1)^{3/2}}$$

The hyperboloid of revolution does not have zero mean curvature, and hence, is not a minimal surface.

---

[68]Monge, *Application de L'Analyse à la Géométrie* 1807, p. 184. Also see Fomenko, 1991, p. 28.

[69]Fomenko, 1991, p. 193. Scherk's minimal surface was derived in 1831, Catalan's surface in 1858, Enneper's surface in 1864 (with Weierstrass), Catalan demonstrated that the helicoid is a ruled surface.

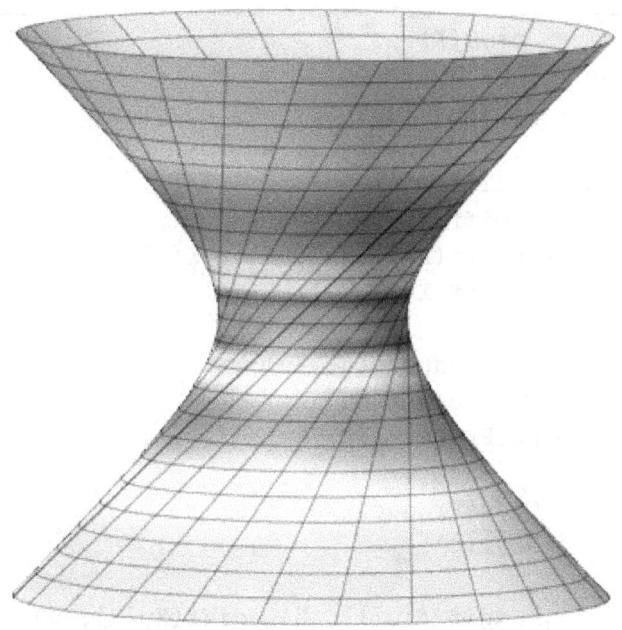

Figure 3.40: Hyperboloid of revolution shaded by its mean curvature: $H \neq 0$

```
MeanH := proc(X)
   local Xu, Xv, E, F, G, n, N,
         Xuu, Xuv, Xvv, e, f, g;

   Xu := <diff(X[1], u), diff(X[2], u),
          diff(X[3], u)>;
   Xv := <diff(X[1], v), diff(X[2], v),
          diff(X[3], v)>;

   E := DotProduct(Xu, Xu);
   F := DotProduct(Xu, Xv);
   G := DotProduct(Xv, Xv);

   n := CrossProduct(Xu, Xv);
   N := Norm(n);

   Xuu := <diff(Xu[1], u), diff(Xu[2], u),
           diff(Xu[3], u)>;
   Xuv := <diff(Xu[1], v), diff(Xu[2], v),
           diff(Xu[3], v)>;
   Xvv := <diff(Xv[1], v), diff(Xv[2], v),
           diff(Xv[3], v)>;

   e := DotProduct(Xuu, n/N);
   f := DotProduct(Xuv, n/N);
   g := DotProduct(Xvv, n/N);

   simplify((G*e+E*g-2*F*f)/(2*E*G-2*F^2));
end:
```

Listing 3.24: Computation of mean curvature, H

```
hyperboloid := <cos(u)-v*sin(u), sin(u)+v*cos(u), v>

MeanH(hyperboloid)

plot3d(hyperboloid,
    u = 0 .. 2*Pi, v = -3 .. 3,
    color = MeanH(hyperboloid),
    style = surfacewireframe, lightmodel = light4,
    scaling = constrained, axes = none);
```

Listing 3.25: Mean curvature: hyperboloid of revolution

In comparison, the mean curvature, H, for the catenoid is zero as computed via the Maple procedure in Listing 3.25. The Gaussian curvature, K, for the catenoid is negative. A plot of the catenoid with parametrization given below is shown in Figure 3.41. Thus, the catenoid is a minimal surface of revolution.

$$X(u,v) = (\cosh u \cos v, \cosh u \sin v, u)$$

$$H = 0 \quad \text{and} \quad K < 0$$

```
catenoid := <cosh(u)*cos(v), cosh(u)*sin(v), u>

MeanH(catenoid)
GaussK(catenoid)

plot3d(catenoid,
    u = -.75 .. .75, v = 0 .. 2*Pi,
    color = MeanH(catenoid),
    style = surfacewireframe,
    lightmodel = light1,
    scaling = constrained, axes = none);
```

Listing 3.26: Mean curvature: catenoid

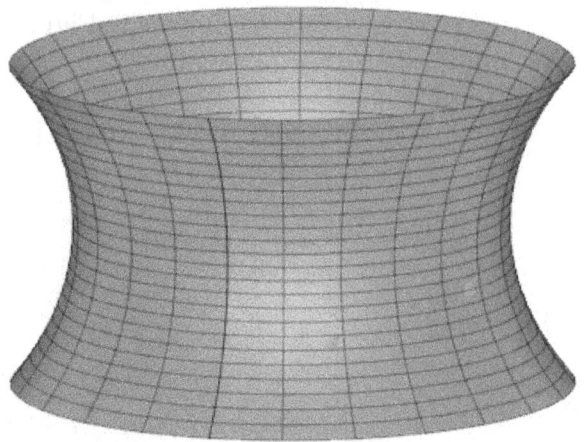

Figure 3.41: Catenoid shaded by mean curvature: $H = 0$

Finally, we consider Enneper's[70] surface given by

$$X(u,v) = \left(u - \frac{u^3}{3} + uv^2, -v + \frac{v^3}{3} - vu^2, u^2 - v^2\right)$$

Studying the results when Listing 3.26 is carried out, we find that Enneper's surface is a minimal surface with

$$H = 0$$

and

$$K < 0.$$

Figure 3.42 shows Enneper's surface with uniform hue, visually identifying it as a minimal surface.

```
Enneper := <u-(1/3)*u^3+u*v^2,
           -v+(1/3)*v^3-v*u^2,
           u^2-v^2>

GaussK(Enneper)
MeanH(Enneper)

plot3d(Enneper,
    u = -2.5 .. 2.5, v = -2.5 .. 2.5,
    color = MeanH(Enneper), style = surfacewireframe,
    lightmodel = light4, scaling = constrained,
    axes = none);
```

Listing 3.27: Mean curvature: Enneper's surface

---

[70]Coddington, 1905, p. 24. Alfred Enneper (1830–1885), in 1868, discovered a new group of self-intersecting surfaces with constant curvature. Since that time, surfaces having these characteristics are known as Enneper's surfaces.

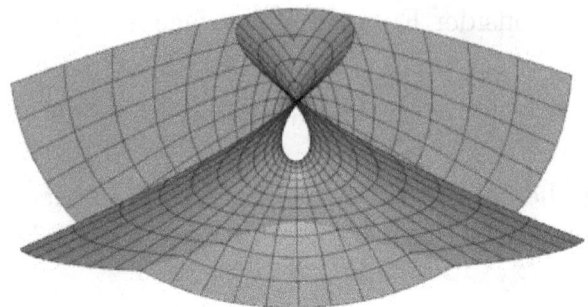

Figure 3.42: Enneper's surface shaded by mean curvature: $H = 0$.

By analyzing these minimal surfaces and their values for H and K, a relationship between the Gaussian curvature and the mean curvature begins to take shape. If $H = 0$, what do we know about K? Given $H = 0$, then $k_1 = -k_2$ or $k_1 = k_2 = 0$. Thus, $K < 0$ or $K = 0$, which tells us that the shape of the surface is either is part of a saddle or part of a plane.

3.8 Monge's Spirit in Differential Geometry

Differential geometry has its roots in calculus, linear algebra, and differential equations. Its development would not have been possible without investigations into map making, accurately measuring time, designing fortifications, surveying, and other technical problems related to geometry.[71] Mathematicians and scientists during the period of the French revolution bolstered and reinvigorated the study of curves and surfaces in space. Monge's steady work in differential geometry led to many advanced applications in the

---

[71] Struik, 1933, p. 191.

subject by a multitude of his students, colleagues, and successors. The modern evolution of differential geometry has progressed to compelling applications in n-dimensional geometry and theoretical physics.

Chapter 4

The Monge Tradition

4.1 The Advancement of Monge's Mathematics by his Students

By far, Monge's greatest success as a teacher is exhibited in the large number of students who developed into strong, independent thinkers because of him. There are numerous works by students of Monge that give testimony to his intellectual creativity and his eloquence as a teacher. At times, these students exceeded their master in one area or another.

> It is a fact that a major part of the epoch-making mathematicians of the early 19th century were former students of adherents of Monge. Even at the end of the century, eminent geometricians, for instance Klein and Lie, still acknowledged their indebtedness to the basic ideas of Monge.[1]

A common characteristic shared by the many mathematicians who were taught by Monge was their connection with useful aspects of

---
[1] Glas, 1986, p.264.

their work. The mathematical lives of two students will be highlighted for the renewed interest in the Monge tradition each of these individuals brought to the subjects of differential and descriptive geometry.

4.1.1 Charles Dupin (1784–1873)

Charles Dupin is considered one of the most important differential geometers among Monge's students.[2] Dupin made many of his mathematical breakthroughs years before they were published in two books, *Développment de géométrie* (1813) and *Applications de géométrie et de mécanique* (1822). The first book was written as a sequel to Monge's textbooks.

When Dupin was sixteen years old, he studied the envelope of the family of spheres tangent to three given spheres. In doing so, he found the *cyclide* (of Dupin), whose families of lines of curvature are circles. His work was published in 1804. In 1807, he proved what came to be known as *Dupin's theorem*; a triple orthogonal system of surfaces intersect in lines of curvature. His work in this area followed Monge's work in interpreting the lines of curvature of an ellipsoid. In studying the curvatures of the normal sections to a surface, Dupin introduced the *indicatrix*. This curve helps us to visualize the behavior of the curvature of a normal section of a surface as the intersecting plane is revolved about the normal. Dupin's approach led to a classification of the points of a surface: *planar, elliptic, hyperbolic,* and *parabolic.* Dupin also clarified the meaning of *umbilic points* and *asymptotic lines.*

Dupin was dedicated to following Monge's methods; without fail he proved each theorem using two different techniques, geometrically and then analytically. In 1819, he published *Essai Historique sur les Services et les Travaux Scientifiques de Gaspard Monge*. In

---

[2]Struik, 1933, p. 117.

this book, Dupin honored Monge as an outstanding man, scientist, and teacher.

4.1.2 Jean-Victor Poncelet (1788–1867)

One particular aspect of Monge's work in descriptive geometry concentrated on representing three dimensional objects in two dimensions by various types of projection. His work led to a revitalized interest in the study of projective geometry in the early 19th century. Whereas Euclidean geometry describes shapes as they are, projective geometry describes objects as they appear. In projective geometry when objects are projected onto another surface, lengths, angles, and being parallel become distorted. The subject originated in the early architectural drawings of the late 14th century.

Jean-Victor Poncelet wrote the first text on projective geometry in 1822, entitled *Traité des propriétiés projectives des figures.* Poncelet was a military engineer who studied at the École Polytechnique. His interest in the problems of projective geometry is directly attributed to the influence of his pedagogue, Monge. In 1812, Poncelet was drafted into Napoleon's army. He was a part of the Russian campaign and was taken prisoner later that year. The two years he spent in Russia enabled him to organize his plans for projective geometry. He often presented his results to his fellow prisoners, some of whom were graduates of the École Polytechnique. A subsection of his 1822 text, "A work of utility for those studying the applications of descriptive and geometric operations on land",[3] shows the influence of Monge's descriptive geometry methods on the development of Poncelet's thought. As an example of his work, Poncelet showed that a conic section is a projective figure. To solve complex problems involving conic sections, Poncelet devised a procedure to simplify the problem by projecting the conic onto a circle, solving the problem on the circle, then per-

---

[3]Kolmogorov, 1981, p.29

forming the inverse projection. In his publication of 1822, Poncelet established the basic concepts, laws, and most significant theorems of projective geometry.

4.2 Monge's Mathematical and Educational Influence in the United States

Monge has gone down in history for his role in the renewal of geometry studies in the nineteenth century and his role in the training of the scientific elite.[4] His successes as teacher, administrator, and founding father of the most famous engineering school during the French Revolution were the reasons why the central subjects of instruction: descriptive geometry, differential geometry, and projective geometry fourished during the first twenty years of the school's existence. Monge's mathematical and educational influence extended far beyond the perimeter of France.

After the poor performance of the United States Army at the beginning of the War of 1812, the administrators of the United States Military Academy at West Point agreed that something could be learned from European practices regarding the training of their military personnel. While Major Sylvanus Thayer was on a mission for the United States Government in 1816, he met Claude Crozet (1790–1864), an engineer and former student of the École Polytechnique. Thayer learned that the military engineers' and officers' educational requirements at the École Polytechnique were based on a strong mathematical curriculum.[5] Using the program of

---

[4]Sakarovitch, 2009, p. 1293.

[5]Kamm, 1998, p. A18. The lasting effect of Monge's administration is reflected in the reputation of the school. According to the article, graduates of the École Polytechnique are considered to be some of the best candidates for the top engineering, political, business, and administrative jobs in France.

study at the École Polytechnique as a model, Thayer's goal was to reconstruct the West Point curriculum. In 1816, Crozet was offered the position of Professor of Engineering at West Point. Crozet followed the standards of his alma mater and introduced Mongean descriptive geometry into the curriculum. In doing so, he became the first teacher of descriptive geometry, a subject of basic importance to engineers, in the United States. In 1821, Crozet published his own textbook, one of the first on the subject written in English, entitled, *A Treatise on Descriptive Geometry for the use of the Cadets of the United States Military Academy*. Claude Crozet became the Chief Engineer for the Commonwealth of Virginia from 1823–1831 and led a successful life thereafter in the United States.

4.3 Conclusion

> Several professors had eagerly gone to listen to their colleague [at the École Polytechnique]. At the end of the session, Monge was surrounded and much congratulated. The compliments spoken by Lagrange had been passed on: 'My dear colleague, what you have just exposed was very elegant; I wish I had done it'. Monge admitted having never received a compliment that touched him so deeply.[6]
>
> René Taton

In 1809, Monge retired from the École Polytechnique and became less active politically, in part due to his diminishing physical health. During this decade, Monge's life would change dramatically. The downfall of Napoleon in 1814 caused the decline of Monge. The commendations he earned were taken away and his

---

[6]Sakarovitch, 2009, p. 1297. Originally quoted in Taton, 1951, p. 216.

administrative offices were left financially in arrears. His fortitude was lost. Monge's wife Catherine Huart, was still regarded as a *comtesse*, however, she had little success in regaining any of her husband's monetary worth.[7] Optimistically, she had twenty copies of Monge's portrait printed from a plate from 1816, with the hope of selling them when necessary.

Monge died on July 28, 1818 in Paris. At his death, he was a man who was well respected by his fellow mathematicians and scientists, and his political adversaries were sympathetic. His loyal follower and friend, Dupin, spoke publicly of Monge's vast achievements at his funeral. Monge's wife expressed to Dupin that she hoped all of her husband's publications would stay in the hands of former students of the École Polytechnique. His life's work made progress in the mathematical sciences possible and his educational practices in technical and cultural subjects affected new generations. A considerable number of his students from the École Polytechnique went out to teach all over France, to impart theoretical and practical training in mathematics and science to a new generation of craftsmen, engineers, and teachers.

The mathematical and educational principles of the founder of descriptive geometry and the father of differential geometry shines on in the 21st century. His development and perfection of descriptive geometry "spread throughout the international scientific community from Europe to the United States."[8] Technical graphical design, in the form of computer models and computer aided design, is not just the present-day form of descriptive geometry. "Probably it is better to say that computer representation systems include descriptive geometry and go over it, as each highly developed sys-

---

[7]Monge married Catherine Huart in 1777. They had three daughters. His wife outlived him and perpetuated his memory in every possible way. She was the only person who stayed close by his side through everything. Even Napoleon would let Monge down in the end.

[8]Cardone, 2003, p. 69.

tem absorbs the previous one".[9] The dissemination of Monge's descriptive geometry laid the foundation for the development of computer-aided design, which validates the importance of the subject in this era.

Monge's creativity and thought provoking work in differential geometry revolutionized the methods mathematicians used in combining geometry and analysis. His processes for determining mathematical concepts, such as curvature, were strikingly different than those of other mathematicians, for example Gauss. Yet, his system of development through differential equations laid the groundwork for the field of differential geometry in its entirety. The *Monge point* of a tetrahedron, the *Monge patch* in surface theory and the *Monge-Ampere equation* are all testaments to his expansive mathematical contributions.

As an educator he was "an exceptional and inspiring teacher, well liked and respected by his students".[10] Monge's contemporaries, family, and friends held him in highest esteem. He was admired for his "quite unusual combination of talents, for he was a capable administrator, an imaginative research mathematician, and an inspiring teacher".[11] Monge's loyalty to his friends, Napoleon, for example, is infamous and gives way to the highly principled person he embodied.

The fundamental role Monge played in the development of mathematics at the turn of the 19th century is well-known and celebrated. The examination of his original ideas and examples, along with their historical and practical motivations, make known another facet of his work that is long lasting — its relevance to the teaching of differential geometry today. One of the novel conclusions drawn is that Monge's work provides an intuitive and pedagogical bridge, via Maple, to illustrate the more difficult theoretical

---

[9]Cardone, 2003, p. 74.
[10]Crabbs, 2003, p. 193.
[11]Boyer, 1960, p. 18.

ideas presented in differential geometry. While not all abstract theorems yield their beauty to graphical computer programs such as Maple, Monge's theorems and the descriptive geometrical methods he used in three dimensions parallel the types of problems for which Maple is inherently suited. Thus, a new tradition in the spirit of Monge may be born in modern differential geometry courses which allows students to explore, discover, and make conjectures from insightful applications. All of this is his legacy.

The illustrious person Monge was and his numerous achievements are commemorated in the underground crypt area of the Panthéon, Chamber VII, in Paris. Monge is buried in Section 18 of the Père Lachaise Cemetery in Paris. It is a rewarding experience to visit Monge's grave and see the monument his students erected in honor of their hero.

Figure 4.1: Monge's Grave at Père Lachaise Cemetery in Paris (front view)

Figure 4.2: Monge's Grave at Père Lachaise Cemetery in Paris (back view)

Appendix A

The Mathematical and Scientific Works of Gaspard Monge[1]

1764 Map of the city of Beaune drawn by Monge and a college friend, Fion, published in *History of the city of Beaune and its Antiques* in Dijon 1772, by abbot Gandelot with the title, "Map of the city of Beaune in 1764 by MM. Monge and Fion"

1766 Two papers on the following subjects, "Mémoire sur la détermination des fonctions arbitraires dans les intégrales de quelques équations aux différences partielles"; "Second mémoire sur le calcul intégral de quelques équations aux différences partielles"

1780 Two papers, "Mémoire sur les fonctions arbitraires continues ou discontinues qui entrent dans les intégrals des équations aux différences finies"; "Mémoire sur les propriétés de plusiers gen-

---

[1] Dupin, 1819 and Taton, 1951. A complete compilation of Monge's contributions to mathematics and science are provided in the given references. A partial listing of his most significant writings from 1764 to 1816 are given in Appendix A.

res de surfaces courbes particulièrernent sur les celles des surfaces développables, avec une Application à la Théorie des Ombres et Pénombres"

1784 "Mémoire sur la théorie des déblais et des remblais"

1785 "Mémoire sur les développées, les rayons de courbure, et les différens genres d'inflexions des courbes à double courbure"

1786 Two papers, "Sur l'expression analytique de la géneration des surface courbes"; "Sur l'intégration de l'équation aux différences partielles dans laquelle les coeficiens L, M N sont quelconques"

1786 "Mémoire sur une méthode d'intégrer les équations aux différences ordinaire, lorsqu'elles sont élevées, & dans les cas ou leurs intégrals complètes sont algebriques"; "Mémoire sur l'intégration des équations aux différences finies ne sont pas linéaires"; "Mémoire sur le résultat de l'inflammation du gaz inflammable et de l'air déphlogistiqué dans les vaisseaux clos"

1788 *Traité élémentaire de statique à l'usage des collèges de la Marine,* Paris. Musier, 227 p., 100 fig. en 5 pl.; "Essai sur le phlogistique et sur la constitution des acides" translated in English by M. Kirwan with notes from MM. de Morveau, Lavoisier, Laplace, Monge, Berthollet, and Fourcroy; "Mémoire sur le fer considéré dans ses différens états métalliques" (with Vandernonde and Berthollet) ; "Mémoire sur l'effet des étincelles électriques, excitées dans l'air fixe"

1789 "Memoire sur quelques effets d'attraction ou de répulsion apparante entre les molécules de matiére"

1791 "Rapport fait a l'Academie le 10 mars, sur le choix d'une unité de mesure" (joint work with Borda, Lagrange, Condorcet)

1793 "Rapport fait à l'Academie le 10 mars, sur le système général des poids et mesures" (joint work with Borda and Lagrange)

1794 *Programme des cours révolutionnaries sur la fabrication de Salpetres, des Poudres, et des Canons* (Monge, Hassenfratz, and

Perrier); *Développements sur l'enseignement l'École Centrale des travaux publics* (with Fourcroy)

1795 *Séances des écoles normales recueillies par des sténographes et revues par des professeurs, texte des lecons de géométrie descriptive données par Monge à l'École Normale; Feuilles d'Analyse appliquée àla Géométrie à l'usage de l'École Polytechnique*

1796 "Sur les lignes de courbure de la surface de l'ellipsoide"

1799 Publication des lecons de l'École Normale on volume séparé par les soins de Hachette: *Géométrie Descriptive;* "Notice sur la fabrication du fromage de Lodézan, connu sous le nom de Parmézan"; Décade égyptienne: "Mémoire sur le phénomène d'optique connu sous le nom mirage"

1802 "Mémoire sur la surface courbe dont toutes les normales sont tangentes à la surface d'une meme sphère"; "Mémoire sur la surface courbe dont toutes les normales sont tangents à d'une meme surface conique à base arbitraire"; *Application d'algebra à la géométrie* (with Hachette)

1805 *Application d'algebra à la géométrie à l'usage de l'École impériale Poly technique* (with Hachette)

1806 "Mémoire sur la surface courbe dont toutes les normales sont tangentes à une meme surface développable quelconque"

1807 *Application de l'analyse à la géométrie*

1808 "Sur quelques propriétés de la pyramide triangulaire"

1810 "Sur les équations différentielles des courbes du second degré"

1816 "Théorème de géométrie"

Appendix B

Courses of Study and Academic Staff at the École Polytechnique (1794–1825)[1]

B.1 Analysis and Mechanics & Descriptive Geometry

Analysis and Mechanics
Ferry (1794–1797)
Lagrange (1794–1798)
Fourier (1795–1796)
Labey (1798–1806)
Lacroix (1799–1808)
Poisson (1806–1814)
Ampère (1809–1827)
Cauchy (1815–1829)
Crozet (1803–1805) (tutor)

---

[1] Dhombres and Dhombres, 1989, Appendix 13. Also, Grattan-Guiness, 2005, pp. 235–236.

Descriptive Geometry
Monge (1794–1809)
Ferry (1794–1797)
Hachette (1810–1812)
Arago (1810–1829)

B.2 Chemistry and Physics

Chemistry
Fourcroy (1794–1808)
Gay-Lussac (1809–1840)
Berthollet (1794–1805)

Physics
Hassanfratz (1794–1814)
Petit (1815–1820)
Dulong (1820–1829)

B.3 Architecture and Fortifications

Architecture
Delorme (1794–1795)
Lomet (1796–1797)
Durand (1796–1833)

Fortifications
Permanent Examiners
Laplace (1796–1799)
Bossut (1796–1807)
Legendre (1800–1815)
Lacroix (1808–1815)
Poisson (1816–1830)

During the first four years of the school's existence, each discipline had their own examiners. After 1797, the school became more stabilized. A partial listing of the examiners is given, note the constancy of the function of the examiner.

B.4 Entrance Examiners

Monge (1798–1807)
Labay (1798–1799, 1806–1814)
Bossut (1796–1807)
Biot (1798–1805)
Dinet (1805–1825)
Reynaud (1809–1825)

Appendix C

Translation of Title Page and *Programme* from *Géométrie Descriptive* by Gaspard Monge, Paris, 1799

DESCRIPTIVE GEOMETRY
―――
LECTURES
TAUGHT AT THE ÉCOLES NORMALES,
Third year of the Republic;
by GASPARD MONGE, of the National Institute
―――
PARIS,
BAUDOUIN, Official Printer for the Legislature and the National Institute
―――
YEAR VII.
―――
PROGRAM

To resolve the problem of the French nation's dependency on foreign industry, which continues to this day, we must first direct our national education toward knowledge about objects that require exactitude, an area that has been totally neglected until now, and to accustom the hands of our artisans to handle instruments of every kind that serve to assure precision in their works and to measure their various dimensions; in this way, our consumers, having become sensitive to exactitude, and our artisans, familiarized with it from the most tender age, will be in a position to achieve it.

Second, we must popularize knowledge about a great number of natural phenomena that are indispensable for industry's progress, and in advancing the general education of the nation, profit from the fortunate circumstance in which the nation now finds itself, having at its disposal the principal resources that it needs.

Finally, we have to disseminate among our artisans knowledge about processes in the arts and about the machinery whose object is to diminish manual labor or imbue the results of the work with greater uniformity and greater precision; and in this regard, one must admit that we have a great deal to learn from foreign countries.

There is no way to fulfill all of these perspectives except to provide a new direction for our national education.

First of all, this would involve familiarizing all intelligent young people, even those with an inherited fortune, with the use of descriptive geometry, so that one day, they will be capable of making better use of their capital both for themselves and for the nation, so that even those persons with no other fortune except their education may be able one day to charge a greater price for their work.

This art has two principal objectives.

The first objective is to represent with exactitude, on drawings with only two dimensions, objects that have three, and which are capable of rigorous definition.

From this point of view, this is a necessary language for any person of genius who conceives a project, for those who must direct its execution, and finally for the artisans who must personally execute its different parts.

The second objective of descriptive geometry is to deduce from an exact description of bodies everything that necessarily follows about their shapes and respective positions. In this sense, it is a method for researching the truth; it offers ceaseless examples of the passage from the known to the unknown; and, since it is always applied to objects capable of the greatest evidence, it is essential to introduce it into the national education. It is not only fitting for exercising the intellectual faculties of a great people and thereby contributing to perfecting the human species, but it is also indispensable for all those workers whose intention is to provide objects with certain predetermined forms; and it is principally because the methods of this art have been so little disseminated until now or even entirely neglected that the progress of our industry has been so slow.

We will thus be contributing to setting the national education in an advantageous direction by familiarizing our young artisans with the application of descriptive geometry to graphic constructions that are required in a great number of arts, and making use of this geometry for the representation and determination of the elements of machinery by which man, taking advantage of natural forces, leaves to himself, so to speak, no other labor in his operations than that of his intelligence.

It is no less advantageous to disseminate knowledge about natural phenomena, which one can turn to the benefit the arts.

The charm associated with these arts may serve to overcome the repugnance people generally have toward strengthening the mind, and to persuade them of the pleasure that comes from using their intelligence, which virtually everyone regards as laborious and tiresome.

Thus, at the école normale[1] there must be a course in descriptive geometry.

But since we have no elementary textbook on this art that is well written, perhaps because scholars have put little interest in it until now, perhaps because it was only practiced in a rather obscure manner by citizens whose education had not been assiduous enough and who did not know how to communicate the results of their inquiries, yet a purely oral course would be entirely ineffective.

Therefore, for the course in descriptive geometry, it is necessary that in addition to learning about methods, there be opportunity for practice and execution.

So those of our citizens whose earlier studies were directed toward geometry or the other exact sciences will have a chance to practice creating the graphic constructions of descriptive geometry in special rooms.

The two parts of this art include general methods, where citizens will familiarize themselves with the use of ruler and compass, without which it would be difficult to imagine their being able to educate themselves.

Among the different applications for which we can use the method of projections, two stand out in terms of their generality and their ingenuity: these are perspective constructions and the rigorous determination of shadows in drawings. These two parts may be considered as complementary to the art of describing objects. We will make certain that the citizens practice these methods because, being destined to teach the procedures of descriptive ge-

---

[1] From the translation of Paul Hamburg: Could be "normal school." These schools were a short-lived contribution of the new Republic to public education, and have the same name as today's elite secondary schools for academically gifted young people in France, but these are not direct descendants.

ometry one day, they need to be familiar with all of its resources. Next, we will apply the method of projections to graphic constructions, which are required in a large number of the arts, such as the practices of stone cutting and carpentry, etc.

Finally, the remainder of the course will be used, first, to describe the basic elements of machines in order to study their forms and effects, and subsequently, to focus on those machines that are most important for general understanding, either because these machines are aimed at making work more precise and uniform or because they are intended to apply the forces of nature to accomplish a particular task, and in this way to augment national power.

Appendix D

Letters and Documents of Gaspard Monge and his family from the D.E. Smith Historical and Gouverneur Morris Collections (1772–1832)[1]

---

[1] Columbia University Rare Books Library, Microfilm #80-1582

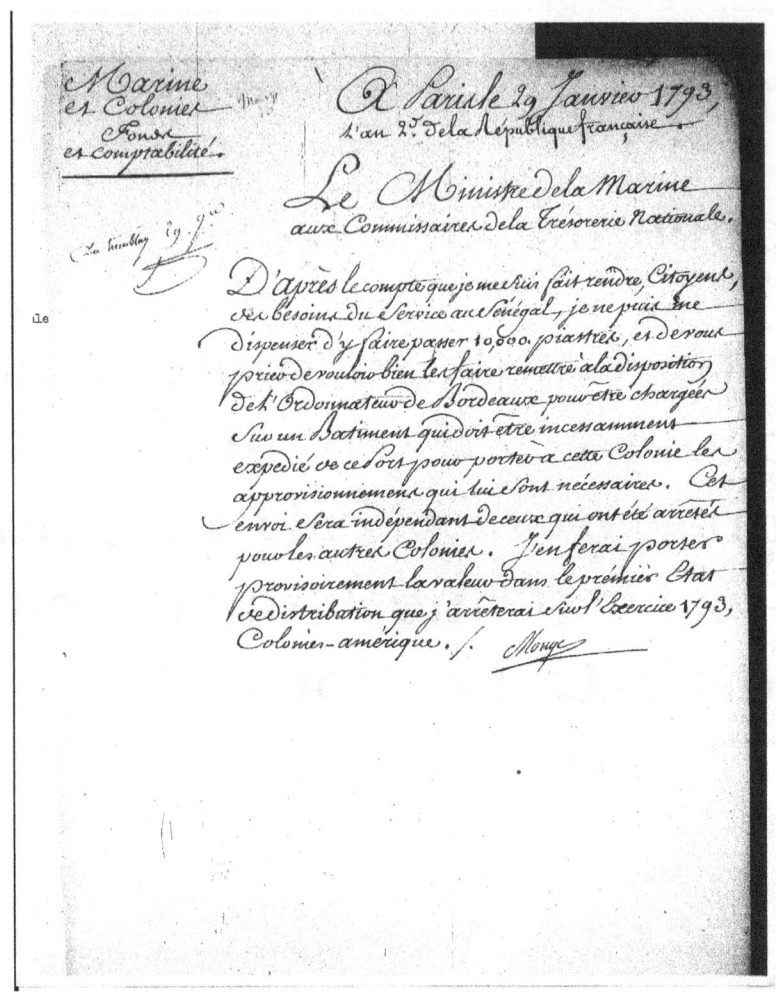

Figure D.1: Letter from Monge in his position as the Minister of the Marine to the Commissary of the National Treasury (1793)

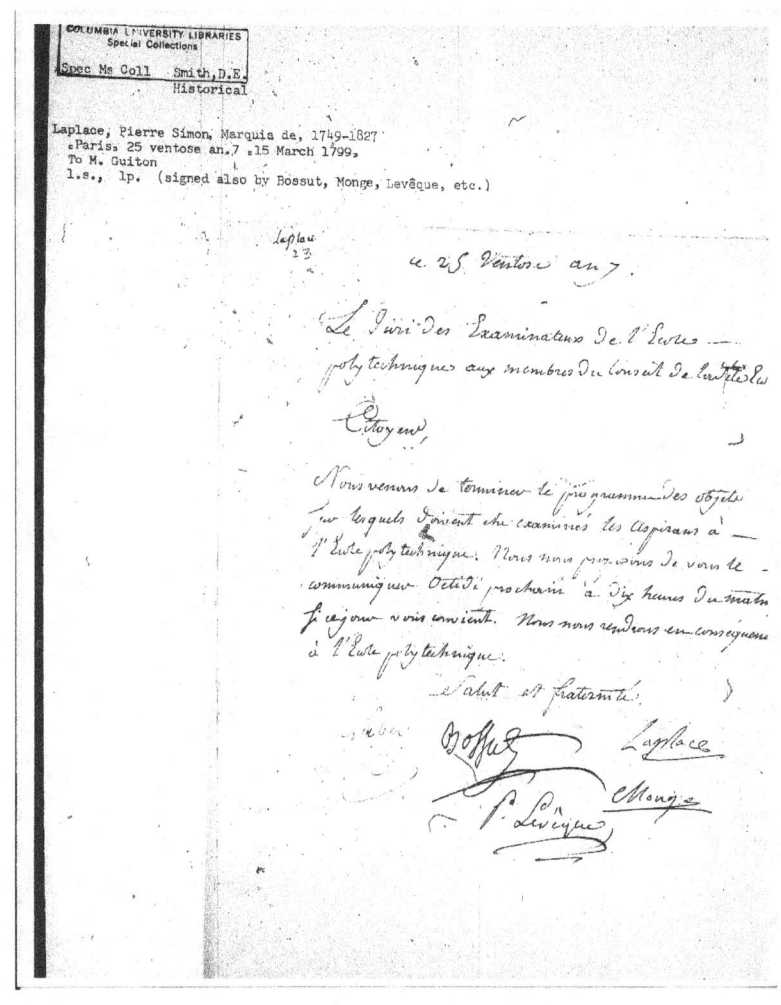

Figure D.2: Letter to Students of the École Polytechnique from the Jury of Examiners: Bossut, Laplace, Monge, and others (1799)

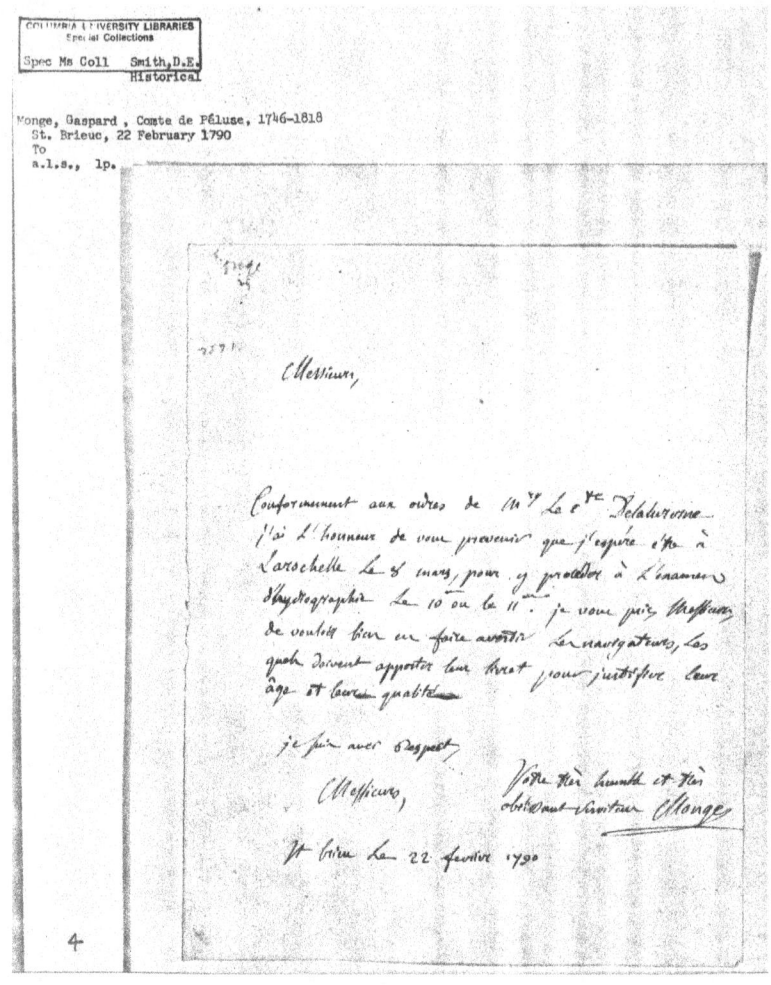

Figure D.3: Letter from Monge in his role as the Minister of the Marine regarding hydrographic testing (1790)

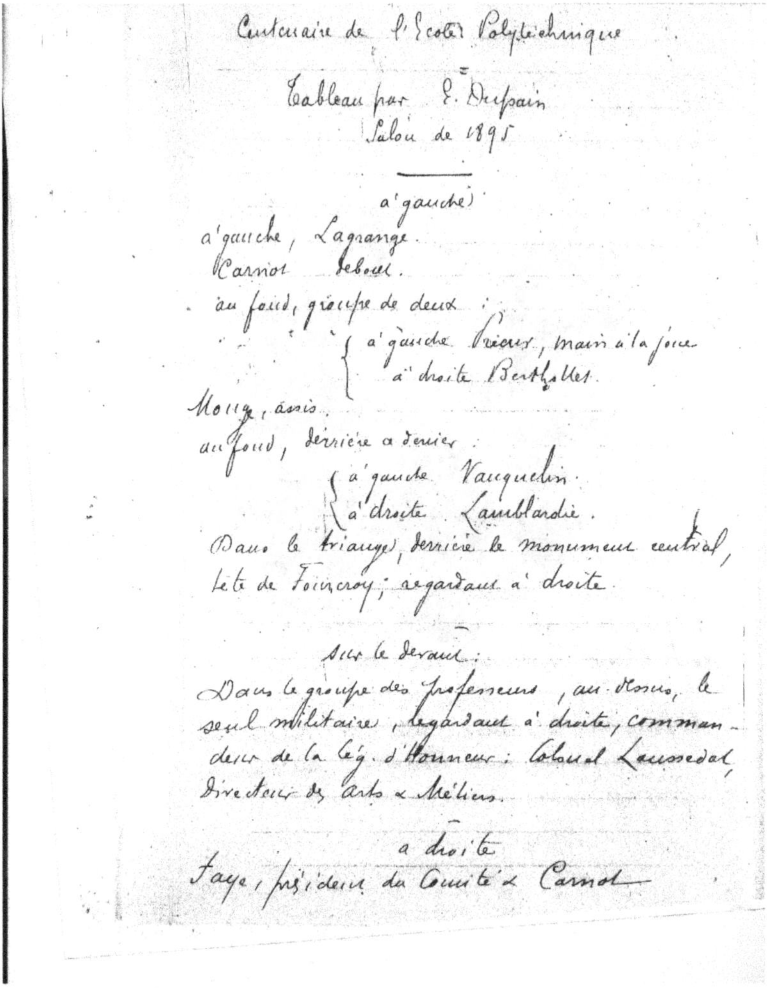
Centenaire de l'École Polytechnique

Tableau par E. Dupain
Salon de 1895

à gauche)
à gauche, Lagrange.
Carnot debout.
au fond, groupe de deux :
{ à gauche Prieur, main à la joue
{ à droite Berthollet.
Monge, assis.
au fond, derrière à demi :
{ à gauche Vauquelin.
{ à droite Lamblardie.
Dans le triangle, derrière le monument central,
tête de Fourcroy, regardant à droite.

Sur le devant :
Dans le groupe des professeurs, au dessus, le
seul militaire, regardant à droite, comman-
deur de la Lég. d'Honneur : Colonel Laussedat,
Directeur des Arts et Métiers.

à droite
Faye, président du Comité et Carnot

Figure D.4: Document regarding the Centennial plans for the École Polytechnique (1895)

Figure D.5: Monge Portrait from Géométrie Descriptive (1799)

Appendix E

Three Translated Letters

E.1   Translated Letter (Figure D.1)

Letter from Monge in his position as the Minister of the Marine to the Commissary of the National Treasury (1793)

> Marine and Colonies
> Funds and Accounting
>
> Paris, 29 January 1793
> Year 2 of the French Republic
>
> The Minister of the Marine
> to the Commissary of the National Treasury
>
> According to the accounting that I have received, Citizen, of the needs of the service in Senegal, I cannot but transfer 10,000 Piasters, and I would ask you to be so kind as to remit this sum for the disposition of the authorizing officer in Bordeaux to be loaded on a boat that will shortly be expedited from there to be carried to this colony for supplying all that is necessary. This allocation will be separate from those budgeted for the other colonies. I will provisionally place the amount in

the first distribution phase that I will budget for the Americas-Colonies 1793.

Monge

E.2 Translated Letter (Figure D.2)

Letter to Students of the Ecole Polytechnique from the Jury of Examiners: Bossut, Laplace, Monge, and others (1799)

25th Ventose (Mid March) Year 7 [1799]

The Jury of Examiners at the École Polytechnique to the members of the of the Council of said school

Citizens,

We have just completed the program of courses for which the candidates at the École Polytechnique must be examined. We would propose communicating this to you next Octidi [Eighth day of the week] at ten o'clock in the morning if this day is suitable for you. If so we will go to the École Polytechnique

I fraternally salute you [standard Republican greeting]

Signatures.

E.3 Translated Letter (Figure D.3)

22 February 1790

Sirs,

Pursuant to the orders of Count [illegible] I am honored to inform you that I hope to be in Larochelle on March 8 in order to perform hydrographic testing on the 10th or the 11th I would ask you, Sir, to be so kind as to alert the sailors, who will need to bring along their brevets to prove their age and their qualifications.

Respectfully yours,

Your very humble and very obedient servant,
Monge

sent on 22 February 1790

Bibliography

Aubry, P. *Monge: le Savant ami de Napoleon Bonaparte, 1746-1818.* Paris, Gauthier-Villars, 1954.

Bell, E. T. *The Development of Mathematics.* The Maple Press Company, York, Pennsylvania, 1940, 1945.

Belhoste, B. "Gaspard Monge", *Les Mathematicians* pp. 50-62, Pour La Science, Paris, 1996.

Belhoste, B. and Dahan-Dalmedico, A., and Dhombres, D., and Laurent, R., and Taton. R.: *L'Ecole Normale de l'An III Lecons de Mathematiques.* Paris Dunod, 1992.

Biemiller, L. "The Serene Grace of Union College's Geometry Models", *The Chronicle of Higher Education,* volume 34 issue 18 (January 13, 1988), p. A3.

Booker, P. J. "Gaspard Monge (1746-1818) and his effect on engineering drawing and technical education", *The Newcomen Society for the Study of the History of Engineering and Technology Transactions* 34: 15-36, 1963.

Boyer, C. B. "Mathematicians of the French Revolution", *Scripta Mathematica* 25: 11-32, 1960.

Cajori, F. *A History of Mathematics.* London, Macmillan & Co., second edition, 1938.

Cardone, V. "From Descriptive Geometry to CAD", in: *Mathematics, Art, Technology and Cinema* pp.67-75, Springer-Verlag, Berlin Heidelberg, 2003.

Casey, J. *Exploring Curvature*. Gottingen, Germany, Hubert & Co, 1996.

Caullery, M. *French Science and its Principal Discoveries since the Seventeenth Century*. Paris, 1933.

Coolidge, J. L. *A History of Conic Sections and Quadric Surfaces*. Oxford, Clarendon Press, 1933.

Coolidge, J. L. *A History of Geometrical Methods*. London, Oxford University Press, 1940.

Crabbs, R. A. "Gaspard Monge and the Monge Point of the Tetrahedron", *Mathematics Magazine* 76: 193-203, 2003.

Crozet, C. *A Treatise on Descriptive Geometry, for the use of the Cadets of the United States Miliatry Academy*, A.T. Goodrich and Co., New York, 1821.

Daston, L. "The Physicalist Tradition in Early 19th Century French Geometry", *Studies in the History and Philosophy of Science* 17 (1986), pp. 269-295.

Dauben, J. "Mathematics in Germany and France in the Early 19th Century: Transmission and Transformation", in: *Epistemological and Social Problems of the Sciences in the Early Nineteenth Century*. Papers from a workshop held at the University of Bielefeld, November 1979, and sponsored by its Institute for the Didactics of Mathematics, eds. H. N. Jahnke and M. Otte, pp. 371-39. Dordrecht, Holland. D. Reidel Publishing Company, 1981.

DeVries, H. "How Analytic Geometry Became a Science", *Scripta Mathematica* 14: 5-15, 1948.

Dhombres, J. and Dhombres, N. *Naissance d'un Nouveau Pouvoir: Sciences et Savants en France 1793-1824*. Paris, Editions Payot, 1989.

DoCarmo, M. P. *Differential Geometry of Curves and Surfaces*. Prentice Hall, Inc, New Jersey, 1976.

Dooley, E. L. *Claudius Crozet French Engineer in America (1706-1864)*. Charlottesville, Virginia, University of Virginia Press, 1989.

Dupin, C. *Essai Historique sur les Services et les Travaux Scientifiques de Gaspard Monge*. Paris, 1819.

Eisenhart, L. P. *A Treatise on Differential Geometry of Curves and Surfaces*. Boston, Ginn and Company, 1909.

Emmer, M. and Manaresi, M. (editors) *Mathematics, Art, Technology, and Cinema*. Springer-Verlag, Berlin Heidelberg, 2003.

Euler, L. "Recherches sur la courbure des surfaces", *Memoires de L'Academie des Sciences de Berlin* [16], pp.119-143 (presented to the Berlin Academy on September 8, 1763). In *Opera Omnia* 1-28, pp. 1-22, 1767. Eneström number E333.

Gafney, L. "Gaspard Monge and Descriptive Geometry," *Mathematics Teacher*, Vol. 58 (Apr., 1965): pp. 338-44.

Gauss, K. F. *General Investigations of Curved Surfaces*. Translated from Latin and German by A. Hiltebeitel and J. Morehead, Introduction by R. Courant. Boston, Ginn and Company, 1909.

Gillispie, C. C. (editor) "Gaspard Monge", In: *Dictionary of Scientific Biography,* volume IX. pp. 469-478. New York Charles Scribner &, Sons, 1971.

Gillispie, C. C. *Science and Polity in France at the End of the Old Regime*. Princeton, New Jersey, Princeton University Press, 1980.

Gilpin, R. *France in the Age of the Scientific State*. Princeton, New Jersey, Princeton University Press, 1968.

Ginsberg, J. and Smith, D. E. *A History of Mathematics in America Before 1900*. Chicago, Illinois, The Mathematical Association of America in cooperation with the Open Court Publishing Company, 1934.

Glas, E. "On the Dynamics of Mathematical Change in the Case of Monge and the French Revolution", *Studies in the History and Philosophy of Science,* I7: 249-268, 1986.

Goursat, E. *A Course in Mathematical Analysis*, Vol. I. Translated by E.R. Hedrick. Boston, Ginn and Company, 1904.

Grant, H. E. *Practical Descriptive Geometry*. New York McGraw-Hill Book Company, Inc., 1952.

Grattan Guinness, I. "Mathematical Physics in France, 1800–1835." In: *Epistemological and Social Problems of the Sciences in the Early Nineteenth Century*. Papers from a workshop held at the University of Bielefeld, November 1979 and sponsored by its Institute for the Didactics of Mathematics. eds. H. N. Jahnke and M. Otte, pp. 349-370. Dordrecht, Holland. D. Reidel Publishing Company. 1981.

Grattan-Guinness, I. "The École Polytechnique, 1794-1850: Differences Over Educational Purpose and Teaching Purposes", *American Mathematical Monthly,* 112 (3): 233–250, 2005.

Graustein, W. C. *Differential Geometry*. New York, Macmillan Company, 1935.

Gray, A. *Modern Differential Geometry of Curves and Surfaces with Mathematica*. Second edition, CRC Press, 1998.

Guggenheimer, H. W. *Differential Geometry*. New York, McGraw-Hill Book Company, Inc., 1963.

Hachette, J. N. P. "Sur les plans osculateurs et les rayons de courbure de lignes planes ou a la double courbure qui resultant de l'intersection de deux surfaces". *Ann. de Maths. pures et appl. de Gergonne,* VII, p. 24-27, 1816.

Hilbert, D. and Cohn-Vossen, S. *Geometry and the Imagination.* New York, Chelsea Publishing Company, 1952.

Hoelscher, R. P. and Jordan, H. H. *Engineering Drawing.* New York, John Wiley & Sons. Inc., third edition, 1946.

Irvine, B. (editor) *Mathematics Teacher,* Vol. 58, pp. 338-344, NCTM, April 1965.

Jomard, E. F. *Souvenirs Sur Gaspard Monge et Ses Rapports Avec Napoleon.* Imprime Par E. Thunot et Cie, Paris, 1853.

Katz, V. J. *A History of Mathematics: An Introduction.* New York, HarperCollins College Publishing, 1993.

Klein, F. *Development of Mathematics in the 19th Century.* Translation of *Vorlesungen über die Entwicklung der Mathematik im 19. Jahrhundert* (1926) by M. Ackerman. Brookline Massachusetts, Math Sci Press, 1979.

Kolmogorov, A. N. and Yushkevich, A. P. (editors) *Mathematics of the 19th Century, Volume II: Geometry, Analytic Function Theory.* Translated from the Russian by Roger Cooke. Boston, Birkhauser Verlag, 1996.

Langins, J. *Conserving the Enlightenment: French Military Engineering from Vauban to the Revolution.* MIT Press, 2004.

Lepage, Jean-Denis G.G. *French Fortifications, 1715-1815: An Illustrated History.* Jefferson, NC, McFarland, 2009.

Le Rouge, Georges-Louis. *Recueil des fortifications, forts,et ports de mer de France.* Le Rouge, Paris, 1755. The Newberry Library, Chicago, IL, Vault Case G1838.L47.

Mahan, D. H. *Descriptive Geometry as applied to the Drawing of Fortification and Stereotomy for the use of the Cadets of the U.S. Military Academy.* First Edition. New York, John Wiley & Sons, 1906.

McCleary, J. *Geometry from a Differentiable Viewpoint.* Cambridge, Cambridge University Press, 1993.

McFarland, J. D. and Rowe, C. E. *Engineering Descriptive Geometry.* New York, D. Van Nostrand Company, Inc., third edition, 1964.

Meusnier, J-B M. "Memoire sur la courbure des surfaces". *Mem. div. sav.* t. 10. Paris, pp. 477-510 (presented to the Paris Academy in February 1776), 1785.

Monge, G. *Application de l'Analyse a la Géométrie.* 1807.

Monge, G. *Application de l'Analyse a la Géométrie.* Reviewed, proofread, and annotated by M. Liouville. Paris, Bachelier, fifth edition, 1850.

Monge, G. *Géométrie Descriptive.* Sceaux, France Editions Jacques Gabay. 1799.

Monge, G. *An Elementary Thesis on Statics: With a Biographical Notice of the Author.* Trans. Woods Baker. Philadelphia, 1851.

Monge, G. *Letters of Gaspard Monge and Family* from the D.E. Smith Historical and Gouverneur Smith Collections (1772–1832). Microfilm MN#80-158, Columbia University Rare Book & Manuscript Library.

Monge, G. "Mémoire sur la théorie des déblais et des remblais", in *Historie de L'Academie Royale des Science, Annee 1781, avec les Mémoires de Mathématique et de Physique.* 1781.

Morrill, W. K. *Analytic Geometry.* Scranton, Pennsylvania, International Textbook Company, 1951.

Morin, A. *Catalogue des Collections du Conservatoire des Arts et Metiers*, Imprimerie de Guiraudet et Jouaust, Paris, 1851.

Olivier, T. *Memoires de Geometrie Descriptive Theorique et Appliquee,* Carillan-Goeury et Vor Dalmont, Paris, 1851.

Oprea, J. *Differential Geometry and its Applications,* Prentice Hall, Inc., Upper Saddle River, New Jersey, 1997.

Parshall, K. H. and Rowe, D. E. *The Emergence of the American Mathematical Community 1876-1900: J. J. Sylvester, Felix Klein, and E. H. Moore.* Providence, Rhode Island, American Mathematical Society, 1994.

Pollak, M. D. *Military Architecture, Cartography and the Representation of the Early Modern European City: A Checklist of Treatises on Fortification in The Newberry Library.* The Newberry Library, Chicago, Illinois, 1991.

Putz, J.F. *Maple Animation.* Chapman & Hall/CRC Press, Boca Raton, Florida, 2003.

Roever, W. H. "Descriptive Geometry and its Merits as a Collegiate as well as Engineering Subject". *The American Mathematical Monthly,* 25: 145-159, 1918.

Roever, W. H. *The Mongean Method of Descriptive Geometry according to the procedure of Gino Loria.* The Macmillan Company, New York, 1933.

Roever, W. H. "Some Frequently Overlooked Mathematical Principles of Descriptive Geometry", *The American Mathematical Monthly* 41: 142-159, 1934.

Roever, W. H. "Some Phases of Descriptive Geometry", *Bulletin of the American Mathematical Society* 31: 540-550, 1925.

Roubaudi, C. and Thybaut, A. *Traite de Géométrie Descriptive classes des Mathématiques Spéciales.* Paris, Masson et Cie Editeurs. eighth edition, 1925.

Sakarovitch, J. "Gaspard Monge Founder of Constructive Geometry", *Proceedings of the Third International Congress on Construction History.* Cottbus, Germany May 2009. Eds. Karl-Eugen Kurrer, Werner Lorenz, and Volker Wetzk. Berlin, Germany: NEUNPLUS1, April 1, 2010. Print.

Salmon, G. *A Treatise on the Analytic Geometry of Three Dimensions.* Cambridge, W. Metcalfe and Son, fourth edition, 1932.

Sarton, G. *The Study of the History of Mathematics.* Cambridge, MA, Harvard University Press, 1936.

Scharlau, W. "The Origins of Pure Mathematics", In: *Epistemological and Social Problems of the Sciences in the Early Nineteenth Century* Papers from a workshop held at the University of Bielefeld, November 1979, and sponsored by its Institute for the Didactics of Mathematics, eds. H. N. Jahnke and M. Otte, pp. 331-347. Dordrecht, Holland, D. Reidel Publishing Company, 1981.

Shell-Gellasch, A. and Acheson, B. "Geometric String Models of Descriptive Geometry", in *Hands on History: A Resource for Teaching Mathematics,* ed. Amy Shell-Gellasch, p. 49-62, Mathematical Association of America, 2007.

Smith, D. E. "Among my Autographs; Monge the Lesser", *The American Mathematical Monthly*, 28: 207-209, 1921.

Smith, D. E. "Gaspard Monge, Politician", In: *Scripta Mathematica* 1: 111-122, 1932.

Stachel, H. "What is Descriptive Geometry For?", *DSG-CK Dresden Symposium Geometrie: konstruktiv & kinematisch*, pages 327-336, TU Dresden, 2003, ISBN 3-86005-394-9.

Stoker, J. J. *Differential Geometry.* New York, Wiley-Interscience, 1969.

Stone, W. C. "The Olivier Models", *The Third Publication of the Friends of the Union College Library,* Schenectady, New York, 1969.

Stone, W. C. "Old Science, new art", *Union College* Volume 75, Number 3, (January February 1983), pp. 10-11.

Struik, D. J. *A Concise History of Mathematics.* New York, Dover Publications Inc., second edition, 1967.

Struik, D. J. *A Source Book in Mathematics, 1200-1800.* Cambridge, Massachusetts, Addison Wesley Press, 1950.

Struik, D. J. *Lectures on Classical Differential Geometry.* Cambridge, Massachusetts, Harvard University Press, 1969.

Struik, D. J. "Outline of a History of Differential Geometry", *Isis* 19: 92-120, 20: 161-191, 1933.

Swetz, F.J. *From Five Fingers to Infinity: A Journey Through the History of Mathematics.* Open Court Publishing Company, Peru, Illinois, 1994.

Taton, R. *L'Histoire de la Géométrie Descriptive.* Histoire Des Sciences, Universite de Paris, 1954

Taton, R. *L'Oeuvre Scientifique de Monge.* Paris, Sainte Germain Presses Universitaires de France, 1951.

Taton, R. and Monge, G. "Un Texte Inedit de Monge: Reflexions sur les Equations aux Differences Partielles". *Osiris* 9: 44-61, 1950. *JSTOR* Web. 7 August 2013.

Weiss, J. "Bridges and Barriers: Narrowing Access and Changing Structure in the French Engineering Profession, 1800-1850", in: *Professions and the French State. 1700-1900* ed. G.Geison, pp. 15-65. Philadelphia, University of Pennsylvania Press, 1984.

Weller, S. W. "Napoleon Bonaparte, French Scientists, Chemical Equilibrium, and Mass Action", *Bulletin for the History of Chemistry* 24: 61-65, 1999.

Willmore, T. J. *An Introduction to Differential Geometry,* Oxford. The Clarendon Press, 1959.